# A Primer on Theory and Operation of Linear Accelerators in Radiation Therapy

**THIRD EDITION**



C. J. Karzmark, Ph. D.*
Professor Emeritus
Department of Radiation Oncology
Stanford University School of Medicine
Stanford, CA

&

Robert J. Morton, M.S.
President
Quality and Regulatory Services, Inc.
Lincoln, CA

Edited By:
James Lamb, Ph. D.
Assistant Professor
Department of Radiation Oncology
David Geffen School of Medicine at UCLA
Los Angeles, CA

*deceased

mpp MEDICAL PHYSICS PUBLISHING

Originally published by the U.S. Bureau of Radiological Health, 1981.
Published as *A Primer on Theory and Operation of Linear Accelerators in Radiation Therapy* by Medical Physics Publishing.
First edition, 1989
Second edition, 1998

Third edition, 2017
ISBN soft cover: 978-1-930524-96-5
ISBN eBook: 978-1-930524-97-2
Library of Congress Control Number: 2017963194



Medical Physics Publishing, Inc.
4555 Helgesen Drive
Madison, WI 53718
www.medicalphysics.org
mpp@medicalphysics.org
(800) 442-5778

Printed in the United States of America

# Contents

# Preface

Welcome to the third edition of the classic book *A Primer on Theory and Operation of Linear Accelerators in Radiation Therapy*. Sadly, Dr. Clarence J. Karzmark is no longer with us, but Robert Morton has enlisted James Lamb to help him update this time-tested book for a new generation studying radiation therapy, radiological physics, radiation oncology, and radiation control.

Electron linear accelerators evolved from the microwave radar developments of World War II. The klystron tube, invented at Stanford University, provided a vital source of microwave power for radar then, as it does now. In the late 1940s, the high-power klystron and the microwave principles incorporated into its design were used to construct and power an electron linear accelerator, or linac, for use in physics research and later for industrial radiography. By the mid-1950s, a linac suitable for treating deep-seated tumors was built in the Stanford Microwave Laboratory and installed at Stanford Hospital, which was located in San Francisco at that time. It served as a prototype for commercial units that were built later.

Since that time, medical linear accelerators have gained in popularity as major radiation therapy devices, but few basic training materials on their operation had been produced for use by medical professionals. Dr. Karzmark, a radiological physicist at Stanford University, was involved with medical linacs since their development, and he agreed to collaborate with Robert Morton of the Center for Devices and Radiological Health (formerly the Bureau of Radiological Health) at the U.S. Food and Drug Administration to write the first edition of this primer. The first primer was originally published by the U.S. Department of Health and Human Services in December 1981 as FDA 82-8181. It provided an overview of a linear accelerator's components and how they function and interrelate. The auxiliary systems necessary to maintain the operation of the linear accelerator were also described. The primer promoted an understanding of the safe and effective use of these devices. It was produced in cooperation with the Division of Resources, Centers, and Community Activities of the National Cancer Institute, and it was intended for students of radiation therapy, radiological physics, radiation oncology, and radiation control.

For ease in understanding, much of this text describes the components as they appear in just one electron linear accelerator treatment unit: the Varian Clinac 18. This choice in no way constitutes an endorsement of this manufacturer's models. Variations in design do occur with other manufacturers and models.

This third edition takes into account the significant advances occurring in radiotherapy linacs since the second edition was published in 1998. Again, the level of treating these advances is simplified so that the audience of radiation therapists—as well as physicians, engineers, and physicists—can benefit. New sections 12 and 13 have been added to describe the essential components of image-guided radiotherapy and physiologic beam gating, respectively, both of which have become standard radiotherapy techniques since the publication of the second edition. The previous section 12 describing beam stoppers has been removed, as these are no longer widely used.

Throughout the text, changes have been made to reflect a modern radiotherapy work flow that is integrated with image guidance and record and verify systems.

# Acknowledgments

The need to simplify complex microwave and physics phenomena while retaining rigor in the treatment of these phenomena presented a significant dilemma in writing this primer. We are deeply indebted to our many colleagues who gave generously of their time in critically reviewing the manuscript, suggesting changes, simplifying analogies, and identifying areas that were unclear. Their incisive comments enabled us to have a better perception of how the primer should be written.

We also wish to acknowledge the assistance, critical review, and encouragement of Bureau of Radiological Health staff members Frank Kearly and Marcia Shane. We thank Craig Nunan for important contributions, in particular, Section 9. For editorial contributions, we acknowledge John R. Cameron, Ph.D.

The original work was supported in part by Research Grant CA-05838 from the National Cancer Institute, NIH, and in part by an Interagency Agreement with the National Cancer Institute, NCI 2Y01-10606.

# 1 INTRODUCTION

Cancer patients are treated by radiation, surgery, or systemic therapies such as chemotherapy. A treatment method proving increasingly effective is radiation, used by itself or in combination with other modalities. The principal radiation modality for the treatment of deep-seated tumors is x-rays of very high energy and penetrating power. Such x-rays are created when high-energy electrons are stopped in a target material such as tungsten. Alternatively, the electrons themselves may be used directly to treat more superficial cancers. The electron linear accelerator (linac) accelerates charged particles in a straight line, in contrast to the circular or racetrack orbits that characterize the cyclotron and synchrotron. The purpose of this primer is to explain the principles of operation and use of the electron linear accelerator and to acquaint the reader with pertinent features and terminology.

The medical linear accelerator will be introduced by first examining the treatment room. Figure 1 shows a patient being readied for treatment with a linac. The thick concrete walls of the treatment room shield the radiation therapist and other staff from the penetrating radiation during treatment. The linac is mounted in a gantry that rotates on a stand containing electronic and other

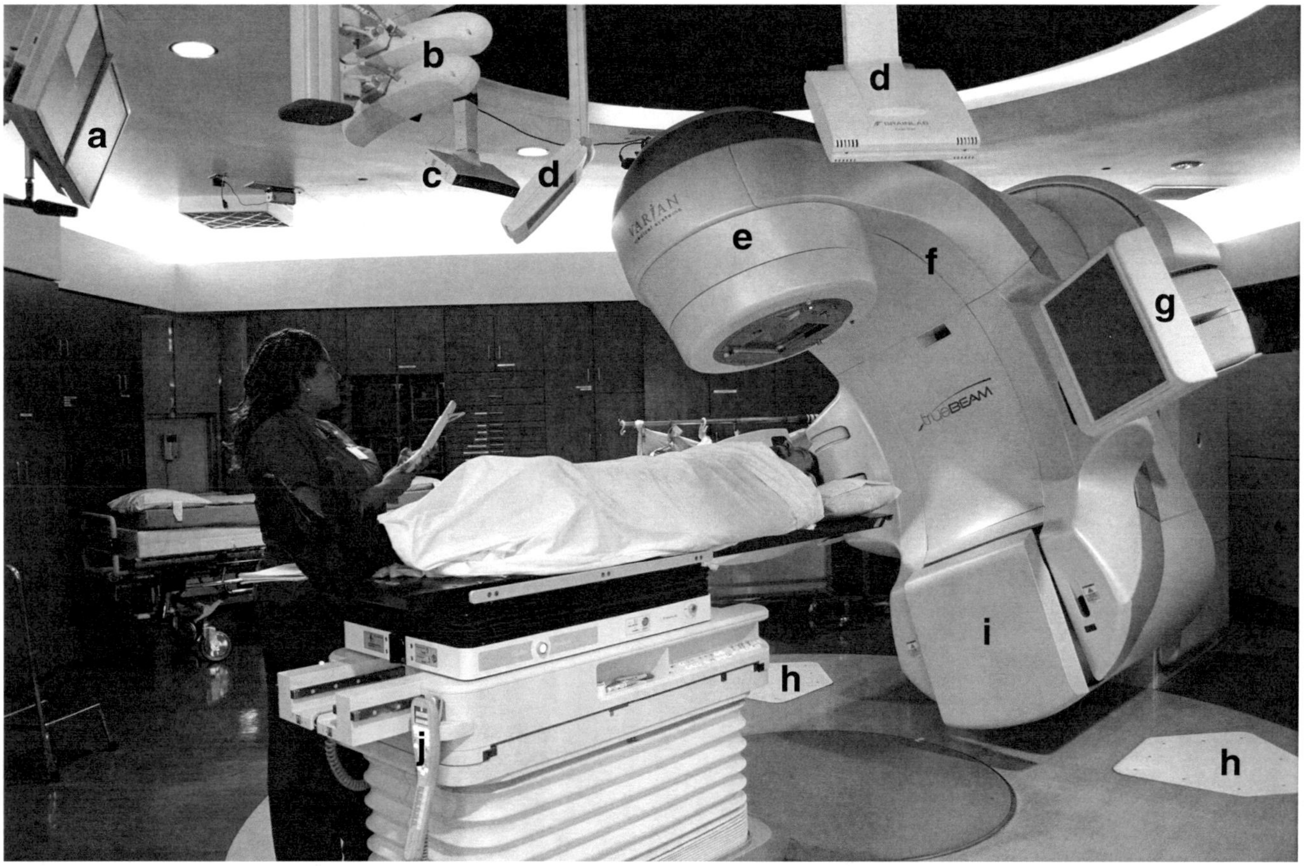


Figure 1. A typical radiation therapy treatment room composed of a Varian TrueBeam linear accelerator and various third-party devices to suit the facilities' needs. a) Position-monitoring flat panel display; b) infrared camera for image guidance (Brain Lab ExacTrac); c) infrared camera (Vision RT); d) flat panel kV detectors (Brain Lab ExacTrac); e) treatment head; f) gantry; g) on-board flat panel kV detector (partially retracted); h) kV x-ray tubes (ExacTrac) under floor panel; i) on-board flat panel MV detector (fully retracted); j) motion control pendant. (Photo courtesy of James Lamb, UCLA.)

systems (Figure 2). The linac can be rotated into position about the horizontal gantry axis for use in treatment. The radiation beam emerging from the collimator is always directed through and centered on the gantry axis. The beam central axis intersects the gantry axis at a point in space called the isocenter. In the majority of cases, the couch is positioned so that the patient's tumor is centered at the isocenter. Usually, the patient lies supine or prone on the treatment couch (sometimes called patient support assembly). The couch incorporates three linear motions and rotation about one or more axes through the isocenter to facilitate positioning the patient for treatment. Side and ceiling lasers project small dots or lines that intersect at the isocenter. These facilitate positioning the patient in conjunction with reference marks, often tattoos, placed on the patient's skin. In a process called image-guided radiotherapy (IGRT), patient positioning is verified and refined using x-ray, optical, ultrasound, or magnetic resonance based in-room imaging systems. Digital position indicators display the treatment field size together with collimator and gantry rotation angles. The isocentric system facilitates comfortable, precise reproducible treatment when using multiple fields directed at the tumor from different gantry angles (Figure 3). In this unit, a constant radiation source-gantry axis-distance (SAD), usually 100 centimeters (cm), is employed. Alternatively,

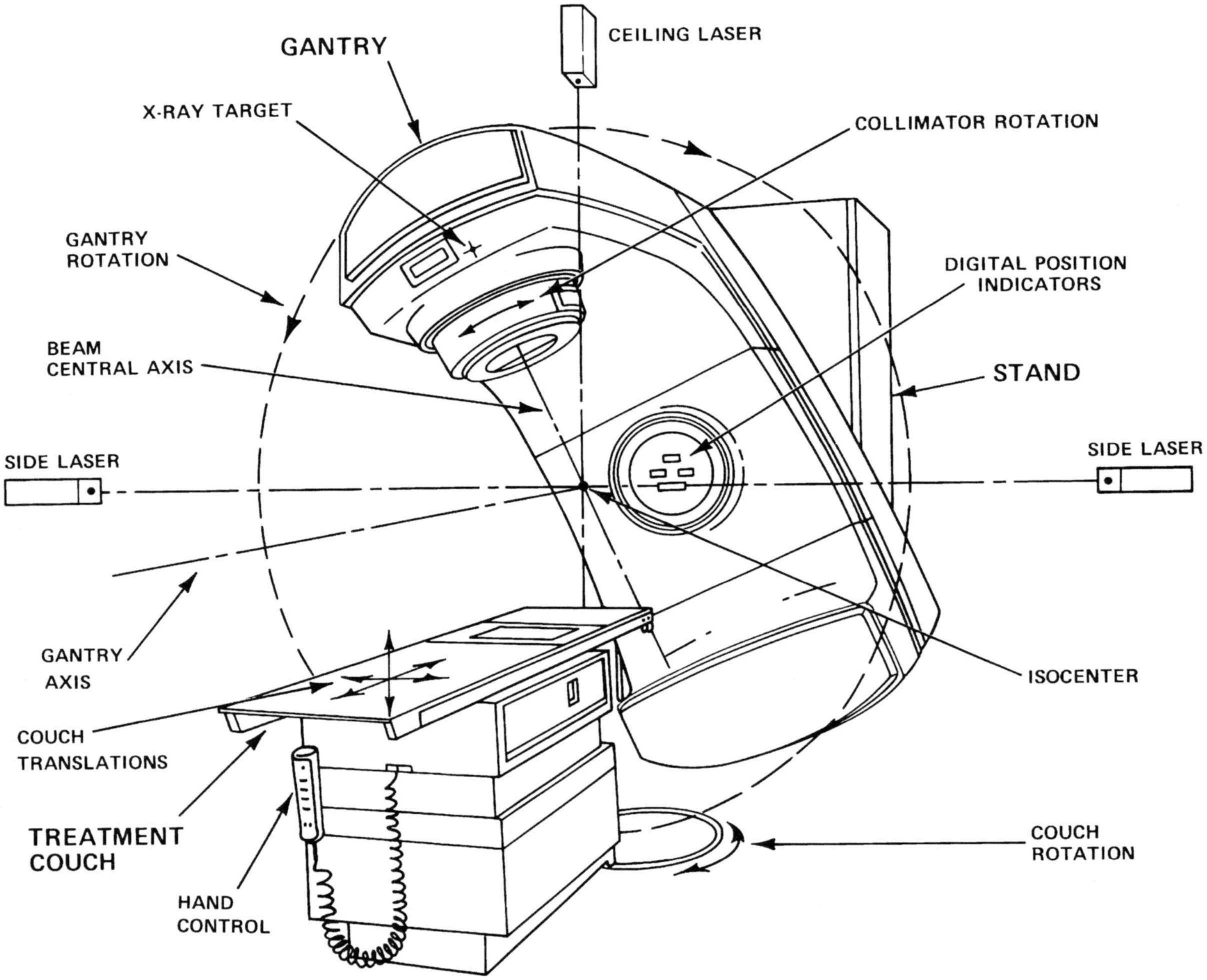


Figure 2. Schematic view of a Varian Clinac 18 linear accelerator system, emphasizing the geometric relationship of the linear accelerator system, positioning lasers, treatment couch, and related motions.

some treatment techniques use a constant radiation source-skin (of patient) distance (SSD), usually for electron treatments or for large fields at distances of 100 cm or greater.

After positioning the patient for treatment in the treatment room, the radiation therapist confirms the treatment parameters from the record and verify system and loads the patient's radiation plan using the control console keyboard (Figure 4). The treatment plan contains collimator positions and monitor units for all the beams of the patient's treatment. From this position at the control console outside the treatment room, the radiation therapist can view the patient on the video monitors and the real-time image of the treatment field on the console monitor display

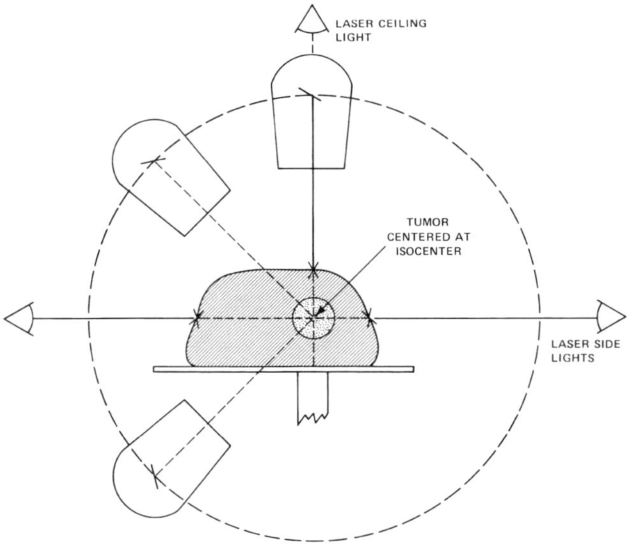


**Figure 3.** The "isocentric" treatment technique. The tumor center, shown within a patient's cross section, is positioned at the isocenter with the aid of skin marks and the lasers shown. The tumor is now positioned for easy and accurate irradiation from any desired gantry angle. The dashed circle depicts all possible x-ray source locations at 100 cm radius (source-axis distance [SAD] = 100 cm).

**Figure 4.** Therapists initiate, monitor, and control the treatment at the control console. Video monitors display two views of the patient and the linac. The record and verify system stores, verifies, and displays the patient treatment parameters. The electronic portal imaging terminal displays a real-time image of the photon treatment field. (Photo courtesy of James Lamb, UCLA.)

and can take emergency action, should it be necessary.

The discussion and illustrations, which follow a brief description of the linac, explain the basic concepts of operation and extend them to the design of an elementary electron linear accelerator. Later, the major modules of a medical linac are identified. Their principles of operation and how they function collectively to produce x-ray and electron treatment beams are described. First, however, we must explain how the energy of a radiotherapy beam is designated.

# 2 ENERGY DESIGNATION IN ACCELERATORS

Figure 5 shows a simple device that will accelerate electrons. It consists of a 1-volt (V) battery connected to two conducting plates spaced 1 cm apart in an evacuated glass tube. The glass tube is an electrical insulator. The negative plate is termed the **cathode** and the positive plate the **anode**. In order to set up the associated electrical charges, the battery causes electrons to flow from the anode to the cathode via the external circuit. This results in a deficiency of electrons at the anode (positive charge) and an excess of electrons at the cathode (negative charge) as shown. This charge distribution creates an **electric "E" field** (denoted by an arrow) in the region between the plates in the direction shown. The size of the electric field is the force that a unit positive charge would feel if placed between the two plates and, in this example, is 1 volt per cm (V/cm). That is, the difference in the electrical potential between the plates, divided by the distance between them, is 1 V/cm. By definition, the arrow identifies the direction a positively charged particle would move; an electron with its negative charge would move in the opposite direction. It is not possible to see "E" fields, but they are known to exist because of the force they exert on charged particles such as electrons. If the electrons in Figure 6 are released from the negative plate (the cathode), they will be accelerated by the force of the "E" field to the positive plate (the anode). An **electron volt** (eV) is the energy gained by an electron accelerated across a potential difference of 1 V.

Exerting a force through a distance is a basic measure of work and energy. On the atomic scale,

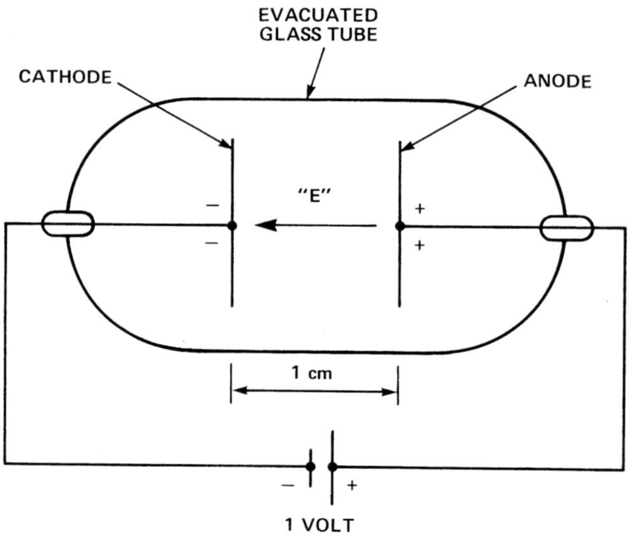


Figure 5. Illustration of an electric "E" field used to accelerate charged particles such as electrons. The "E" field is directed to the left, i.e., from the positive to the negative terminal.

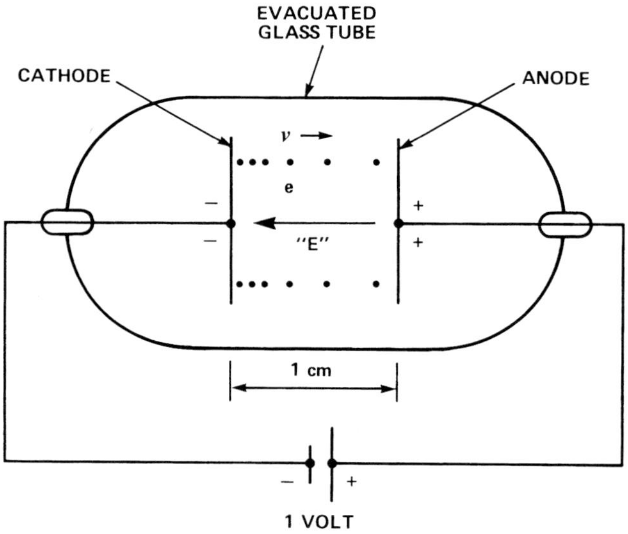


Figure 6. A simple electron linear accelerator of energy one electron volt (1 eV). Electrons, "e," are depicted by dots moving with a velocity ($v$) to the right. The "E" field is one volt per cm (1 V/cm) in the opposite direction. Note that the density of electrons is highest near the cathode, where their velocity is lowest.

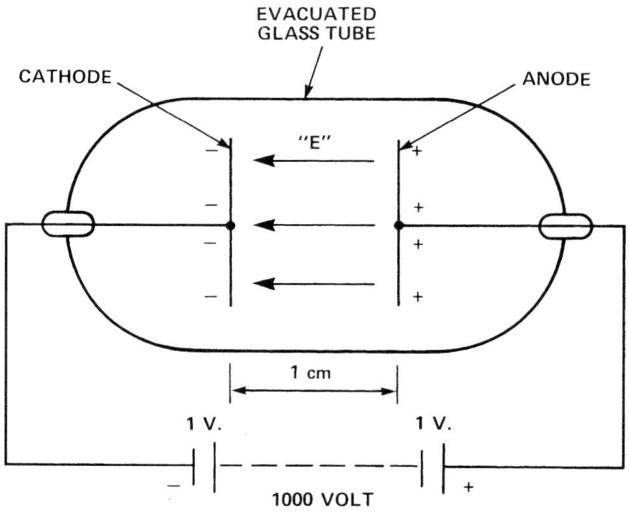


Figure 7. A simple electron linear accelerator of energy 1000 electron volts (1 keV).

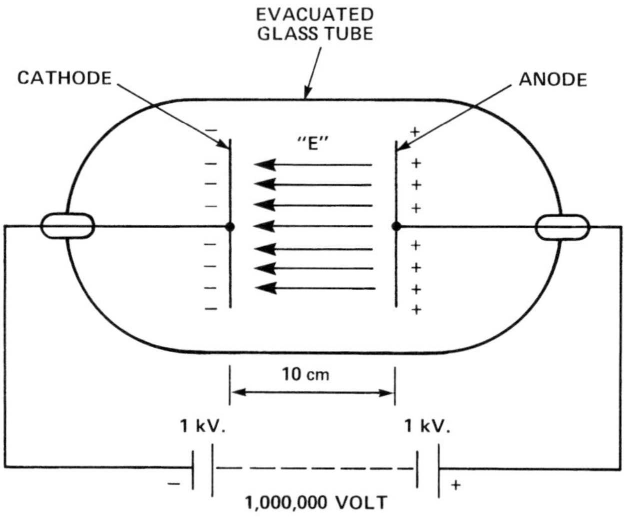


Figure 8. A simple electron linear accelerator of energy one million electron volts (1 MeV). Note the increased electric charge (+, –) distributions at the anode and cathode and the greater "E" fields compared with those for the lower voltages of Figures 5, 6, and 7.

the electron volt, or multiples of it, is the adopted unit of energy. In Figure 6 we are dealing with a force of 1 V/cm exerted on an electron through a distance of 1 cm.

Imagine now that a thousand 1-V batteries are connected in series to provide 1,000 V, or 1 kV potential differential, across the plates of this device as in Figure 7. The accelerated electron would arrive at the anode with an energy of 1000 eV or 1 kiloelectron volt (1,000 eV = l keV). Note also that the strength of the associated "E" field now is 1,000 volts per cm (1 kV/cm).

Suppose the plates are spaced 10 cm apart and that a thousand 1-kV batteries are connected in a series to provide a 1-million-V power supply (Figure 8). The plate spacing and glass tube have been lengthened to withstand this higher voltage without electrical breakdown. An electron released from the cathode now gains 1 million eV of energy during its transit and arrives with an energy of 1 million eV (1,000,000 eV = 1,000 keV = 1 MeV). Note that the energy gained by the electron depends only on the potential difference between the anode and the cathode, and not the distance traveled. The corresponding electric field strength "E" is 1 million V divided by 10 cm, or 100,000 V/cm (100 kV/cm). To establish the higher electric field strengths of Figures 7 and 8, the + and - charge distributions at the anode and cathode are proportionately larger, as compared with Figures 5 and 6. To simplify the figures that follow, the "E" lines and associated charge distributions at the anode and cathode will sometimes be omitted. Linear accelerators give energy to charged particles by accelerating them in a straight line.

# 3 AN ELEMENTARY LINEAR ACCELERATOR

It is possible to convert the simple linac just described to a more sophisticated, yet still elementary, electron linear accelerator. First, a heated cathode is substituted for the negative plate (Figure 9). The cathode shown is a simple filament. (In the linac this cathode becomes the electron gun.) The small battery, B, heats the filament, causing it to literally "boil off" electrons. Next, a theoretical 6-MV battery is connected between the

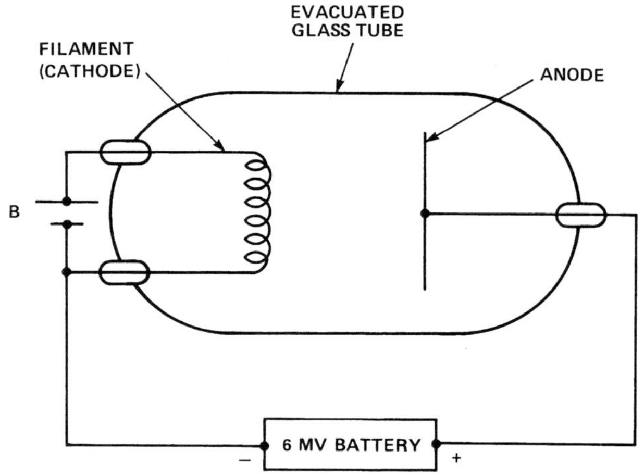


Figure 9. An elementary six million volt (6 MeV) electron linac. The filament-type cathode for the electron source is heated by a small battery, B.

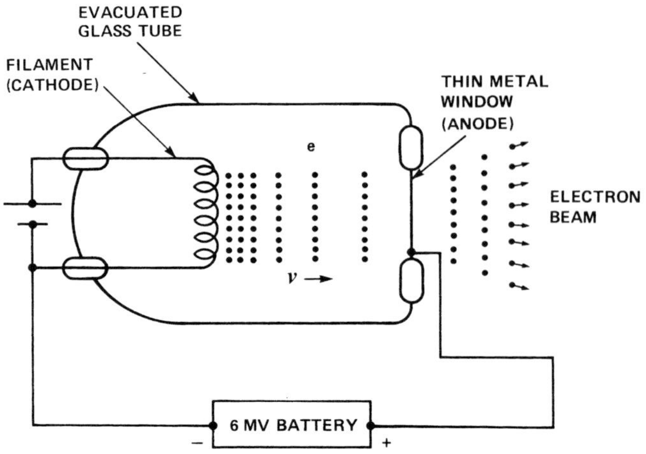


Figure 10. The six-million-volt linac of Figure 9 modified to provide external 6-MeV electrons (as for therapy). Electrons, "e," are accelerated and emitted continuously in this simple, battery-powered linac.

cathode and anode. This battery voltage corresponds to the electron energy desired, i.e., 6 million V for 6-MeV electrons. Here, electrons are boiled off the filament and accelerated to an energy of 6 MeV as they strike the anode.

To adapt this linac for electron therapy, a thin metal "window" becomes the positive plate or anode (Figure 10). Such a thin, solid, metal sheet maintains the necessary vacuum and yet permits the electrons to penetrate the window and emerge with only a small loss of energy. In this elementary linac, the electron beam emerges with an energy only slightly less than 6 MeV.

To adapt this linac for x-ray therapy, the positive anode placed outside the window is a thick tungsten target that stops the electrons abruptly, thereby producing penetrating x-rays (Figure 11). These x-rays will have energies from a fraction of an MeV up to 6 MeV, all initiated by electrons of 6 MeV energy, since the electrons can give up their energy all at once in a single collision or in several sequential collisions. The resulting spectrum of x-ray energies is designated by "6 MV." The notation convention of dropping the "E" from "MeV" indicates that the x-ray beam will be made up of x-rays of different energies produced as the 6-MeV electrons are slowed and stopped in the target.

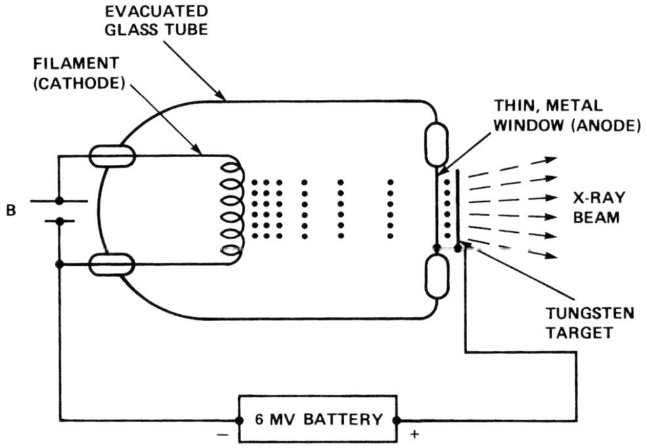


Figure 11. The elementary six-million-volt linac of Figure 10 modified by providing a target to provide external 6-MV x-rays.

Instead of energizing this simple linear accelerator with a battery, substitute an alternating voltage, as shown in Figure 12. The magnitude and polarity of such a voltage changes regularly and repeats itself periodically with time in this cyclic pattern, which is called a sine wave. For the single

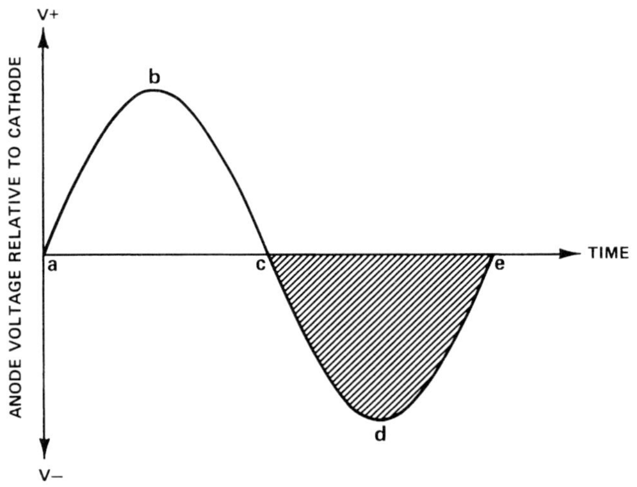


Figure 12. One cycle of the alternating voltage used to power the linac shown in Figures 13 and 14. The anode voltage, V, relative to the cathode is plotted as the ordinate against time, as an abscissa. This pattern repeats itself at the frequency of the alternating voltage (60 cycles per second or 60 Hz for this elementary linac) as time progresses and is called a sine wave.

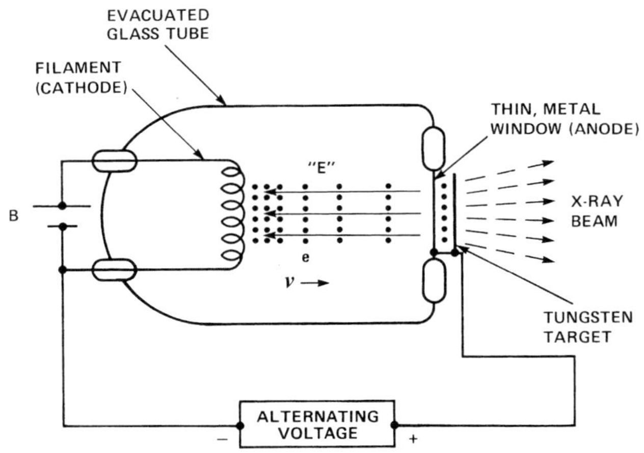


Figure 13. The elementary electron linac for x-ray therapy modified to use an alternating voltage power source, whose polarity reverses every half cycle (see Figure 12). For the polarity shown, electrons will be accelerated to the anode. Note that "E" and the electron velocity ($v$) are in opposite directions.

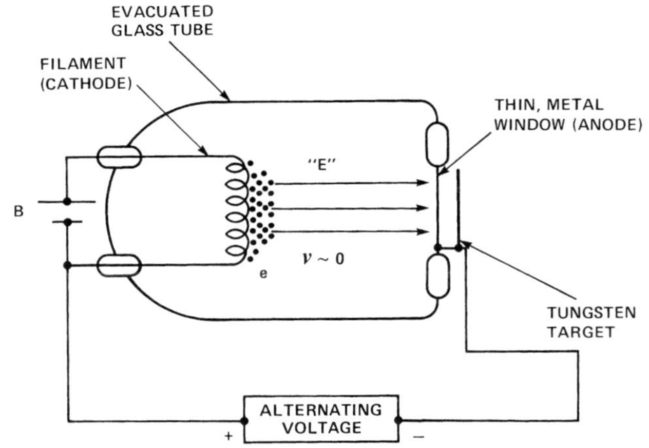


Figure 14. The elementary linac for x-ray therapy powered by an alternating voltage. For the polarity shown (opposite of Figure 13), electrons remain near the filament and will not be accelerated to the anode.

cycle shown in Figure 12, the horizontal axis denotes time; the vertical axis denotes the magnitude and polarity of the anode voltage, V, relative to the cathode, that establishes the "E" field. Many electrical and mechanical phenomena change smoothly in this regular pattern of a sine wave. The number of complete sine wave cycles per second (+ and – for up and down excursions) is called the frequency and is measured in hertz (Hz), kilohertz (kHz), or megahertz (MHz). One hertz equals one cycle per second. Typically, the frequency of home electric power is 60 Hz, a standard (AM) broadcast radio wave can be 1000 kHz, and an FM radio wave can be 100 MHz. The energizing frequency for most medical linacs is 3000 MHz. The latter high frequency is referred to as a microwave frequency. The time for completing a single cycle is called the period and, for the above examples, coincides with 1/60th second, 1/1,000,000th second (1 microsecond), and, lastly, 1/3,000th of a microsecond, respectively.

Most linacs are powered by a 60 Hz line voltage (see Figure 12), whereas the electrons are accelerated by a 3000 MHz voltage. With the target positive and the filament negative, as shown in Figure 13, electrons emitted from the cathode during interval a-b-c of Figure 12 are accelerated to the target. At point c in Figure 12, the voltage reverses polarity and "E" field direction. With the target negative and filament positive, during interval c-d-e as shown in Figure 14, electrons are still emitted and form a "cloud" of electrons around the filament but are not accelerated to the target.

Electrons are accelerated only during the first half of the cycle shown in Figure 12. At a time one-fourth through the cycle, point b, the voltage, V, reaches a positive maximum, and "E" is directed as in Figure 13. An electron released from the cathode at this time would gain a maximum of energy. Conversely, at three-fourths time through the cycle, point d, V reaches a negative maximum. Then, "E" is directed as in Figure 14, and maximum energy would have to be expended in "pushing" the electron from the cathode if it were to reach the anode against the opposing "E" field. Therefore, no electrons are accelerated at this time. At other times, intermediate amounts of energy would be gained or expended, including zero at points a, c, and e. In this illustration, we assume that the electron travels between cathode and anode instantly; that is, the electron's travel time is zero. Note that now the elementary linac accelerates electrons and emits radiation only half of the time, and the electrons vary in energy sinusoidally during this time.

# 4 A COMPARISON OF LINACS WITH DIAGNOSTIC X-RAY GENERATORS

There are many similarities between the linac and a conventional diagnostic x-ray generator. Both provide a source of electrons from a hot filament or cathode in an evacuated tube. Both require an accelerating voltage between the cathode and target anode. This voltage is adjustable in a diagnostic generator, depending on the procedure, from about 30 to 150 kV. In contrast, linac accelerating voltages are fixed in a particular unit and range from about 4 to 35 MV. Diagnostic x-rays often involve a single 0.01- to 10-second pulse with a 60 Hz structure with higher-frequency ripple components, while linac radiation consists of short bursts of about five-millionths of a second duration repeated several hundred times per second, each burst having a 3000 MHz frequency structure. Both employ collimators to shape the x-ray beam, but these must be thicker in the case of linacs. Because of their high energy, x-rays from linacs are much more penetrating than diagnostic x-rays. This is a distinct advantage for treating a deep-lying cancer because the cancer cells can be destroyed by the linac beam with less damage to healthy, overlying tissues.

Linacs require heavily shielded rooms to protect the persons outside. Such rooms are usually constructed with thick concrete walls. In contrast, diagnostic rooms are usually shielded by a sheet of lead a few millimeters thick hidden in the walls. Diagnostic x-rays reveal anatomical structures based on differences in atomic number as well as physical density, e.g., bone versus soft tissue or air; megavoltage x-ray attenuation is primarily based on density differences. A film produced with megavoltage x-rays would show little difference between bone and soft tissue. The importance of x-ray diagnostic beams is primarily in the information contained in the transmitted beam, which produces an image on a receptor. The importance of x-ray therapy beams, such as provided by linacs, is primarily in the energy absorbed in the tumor, not in the beam transmitted through the patient.

Orthovoltage (about 250 kV) radiation equipment, which dominated treatment energies of the 1930s, has properties closer to diagnostic x-rays than megavoltage energy therapy beams and continues to be appropriate for specific treatments.

# 5 MAJOR LINAC MODELS AND COMPONENTS

Contemporary linacs consist of a number of major modules and components that will be identified and their operating principles described. This information will be combined and interrelated to explain an operational linac.

The major modules in the linac are the gantry, the stand, the control console, and the treatment couch (Figures 1 and 4). Some linacs also have an external modulator cabinet, as Figure 42 shows. Figure 15 identifies the components housed in the stand and gantry of a high-energy linac and will be referred to frequently. The stand is anchored

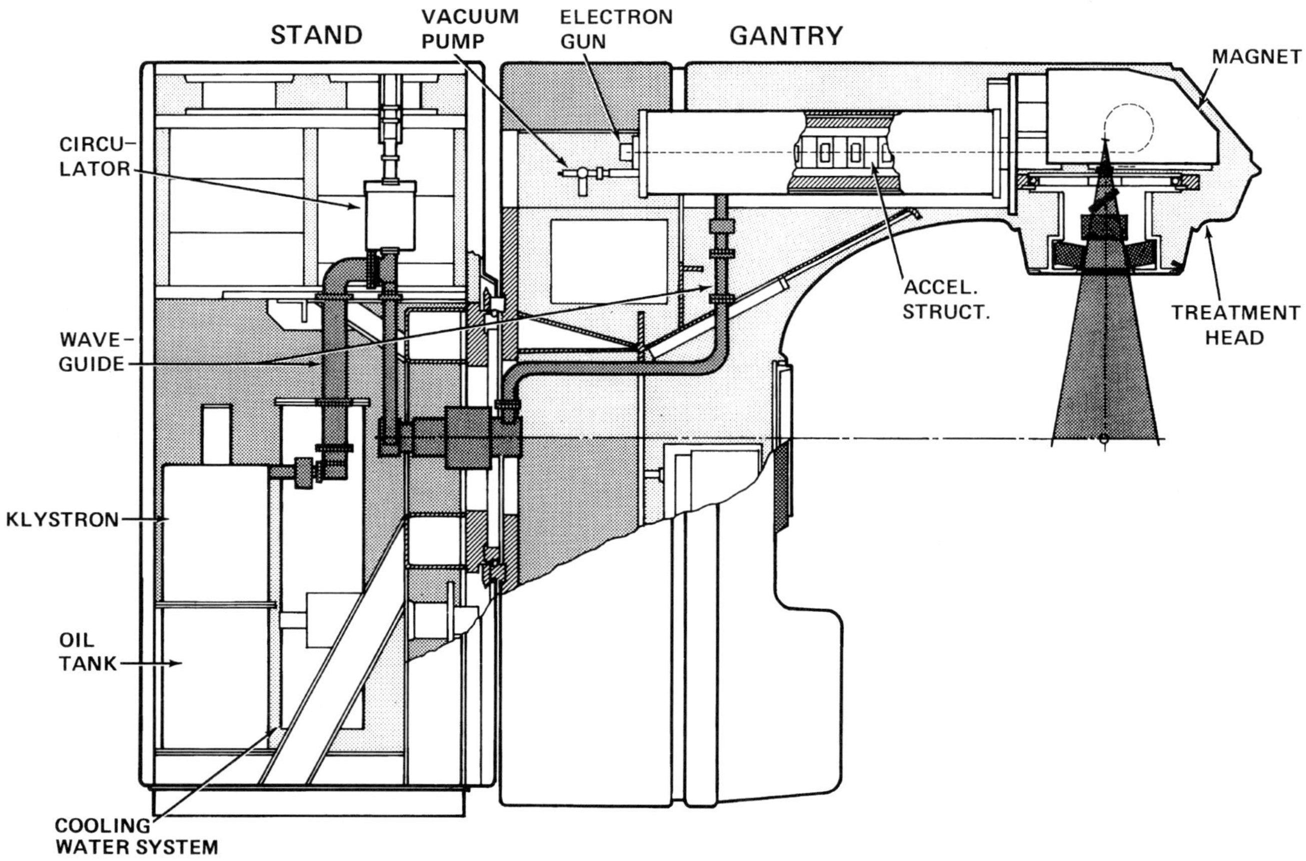


Figure 15. Schematic diagram of a high-energy radiotherapy linac (Clinac 18) identifying major components housed in the stand and gantry.

firmly to the floor, and the gantry rotates on bearings in the stand. The operational accelerator structure, housed in the gantry, rotates about a horizontal axis fixed by the stand. For ease of understanding, most of the text describes the components as they appear in the Varian Clinac 18.

The major components in the stand are the following:

1. **Klystron**—a linear series of microwave cavities that sit atop an insulating oil tank and provides a source of microwave power to accelerate electrons;
2. **Waveguide**—conveys this power to the accelerator structure in the gantry;
3. **Circulator**—a device inserted in the waveguide to isolate the klystron from microwaves reflected back from the accelerator; and
4. **Cooling water system**—cools various components that dissipate energy as heat and establishes a stable operating temperature for the accelerator structure that is sufficiently above room temperature to prevent condensation of moisture from the air.

The major components found in or on the gantry are the following:

1. **Accelerator structure**—a linear series of microwave cavities that are energized by the microwave power supplied from the klystron via the waveguide;
2. **Electron gun** (or cathode)—provides the source of electrons injected into the structure;
3. **Bending magnet**—deflects the electrons emerging from the accelerator structure around a loop focusing the electron beam on the target to produce x-rays or to be used directly for electron treatments; and
4. **Treatment head**—contains beam shaping and monitoring devices.

Additionally, most modern linear accelerator units have image guidance components mounted on the gantry, including an **electronic portal imaging device (EPID)**, an **on-board kV x-ray generator**, and an **imaging flat-panel detector**.

The modulator cabinet (Figure 42) contains components that distribute and control primary electrical power to all areas of the machine from the utility connection. It also supplies high-voltage pulses for beam injection and for generating microwave power.

The treatment couch motions are controlled by a hand pendant control operated by the radiation therapist (Figure 1). The three-dimensional positioning of the patient on the couch is motor-driven. Fast and slow speeds or variable-speed motor control are provided for the couch, together with control of gantry rotation and secondary collimator adjustments (Figure 2). Most couches also provide couch rotation around a vertical axis passing through the isocenter, and some permit attachment of a treatment chair. Also available are "6D" couches, which provide robot-controlled translational motion in three directions and rotation about three axes.

The control console (Figure 4) is the operations center for a linac. It supplies the timing pulse that initiates each pulse of radiation. It provides visual and electronic monitors for a host of linac operating parameters, including the individual patient's treatment dose. The prescription is for the total series of treatments. Treatment cannot proceed when the treatment parameters exceed limits that have been previously established.

In addition to these major modules and components, there are a number of auxiliary systems, including vacuum and water pressure, temperature control, automatic frequency control (AFC), as well as radiation monitor and control (see section 13).

# 6 MICROWAVE POWER SOURCES

The klystron and magnetron are two special types of evacuated electron tubes that are used to provide microwave power to accelerate electrons. Klystrons are used to power high-energy linacs, while magnetrons power lower-energy linacs. Microwaves are similar to ordinary radio waves, but they have frequencies thousands of times higher. The microwave frequency needed for linac operation is typically 3 billion cycles per second (3000 MHz). See Sections 3 and 4. The voltage and "E" fields associated with microwaves change sinusoidally in direction and magnitude in a regular manner, producing an alternating voltage as shown in Figure 12. Microwave cavities, which are central to the construction and operation of klystrons and magnetrons as well as linear accelerator structures, will be described next.

## 6a. Microwave Cavities

Microwave devices—including klystrons, magnetrons, and accelerator structures—make extensive use of resonant microwave cavities. A simple microwave cavity similar to that used in medical linacs, but with closed ends, is shown in Figures 16 and 17. It is an accurately machined cylinder, about 10 cm in diameter and several centimeters in length. Such a cavity has the approximate size and shape of a tuna fish can. In Figure 18, the cavity is shown modified by cutting openings in its two ends along the axis, for use in a klystron or an accelerator structure. A microwave cavity is an enormously efficient device in the sense that the intense "E" fields needed for these applications are established by a small amount of electrical power. This is a resonance phenomenon that occurs at one frequency, in this case 3000 MHz, which is determined by the dimensions of the cavity, much as a musical organ pipe of a particular length resonates to a particular pitch. Such cavities are formed of copper for high electrical and thermal conductivity. An electric current, I, flows on their inner walls, moving electric charge from one cavity end to the other, as shown in Figures 16, 17, and 18. These end regions of dense electric charge are central to both klystron and accelerator structure operation because they give rise to the intense "E" fields along the axis of the cavity, as in Figures 16b, 17b, and 18b. The magnetic "H" field pattern of Figures 16c, 17c, and 18c that exists in the cavity will be omitted in the illustrations that follow, because they are unimportant for our purposes.

The electric and magnetic fields and currents and charge distributions that exist in a cavity have a complex dependence on time. They have been

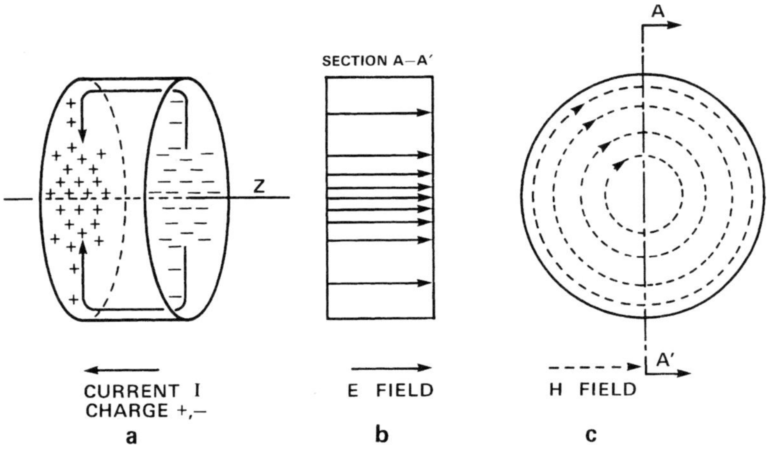


Figure 16. The electric charge (+, –), current I, and electric "E" field and magnetic "H" field patterns for a closed cylindrical microwave cavity. The cavity wall current, I, is circularly symmetric around the Z axis of the cavity. The "E" and "H" fields fill the entire cavity volume. Section b is made by a plane surface cut through the center of the cylinder containing the Z axis. Section c is cut perpendicular to the Z axis.

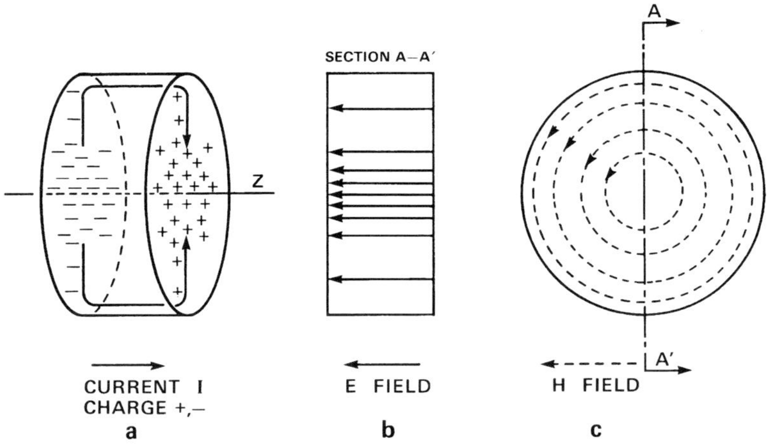


Figure 17. Closed cylindrical microwave cavity with polarity opposite of that shown in Figure 16.

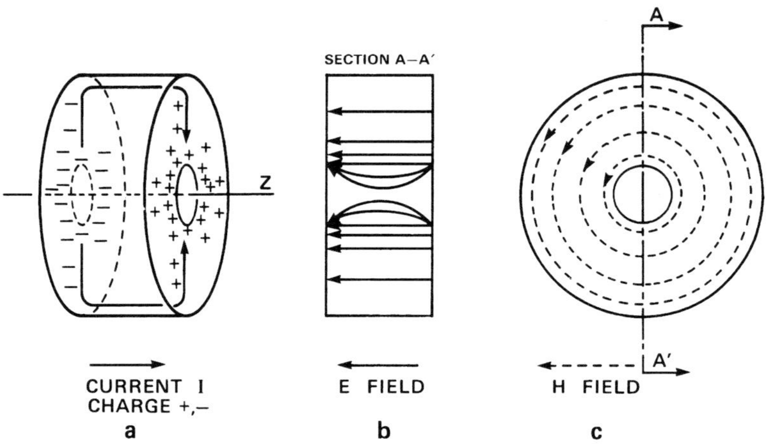


Figure 18. Cylindrical microwave cavity with the same polarity as Figure 17, but with circular holes cut in the ends of the cavity. The holes concentrate the current, I, charge, and the "E" field and facilitate acceleration of electrons along the Z axis.

separated arbitrarily in Figures 16, 17, and 18a–c, for clarity. The polarity of the electric charge and current, and the "E" and "H" field directions reverse twice each microwave cycle, that is, 6 billion times a second! The patterns of Figures 16 and 17 are one-half cycle apart in time. In order to take advantage of these intense "E" fields to build a klystron or an accelerating structure, circular openings on axis at the cavity ends are cut as shown in Figure 18 so that electron beams can be introduced to interact with these fields. The electron beam current passes through these openings along the cylindrical axis Z. The large cavity wall currents, I, should not be confused with the smaller axial electron beam current, which originates from an electron gun in a klystron or in an accelerator structure. The arrows denoting I in Figures 16a, 17a, and 18a point in the direction that a positive charge current would flow. (Note that the axial electron beam current is not shown in these figures.) The negative electrons, which are the charge carriers in the pulse, flow in the opposite direction.

Earlier, the energy transfer from a static, and then an alternating, electric "E" field to an electron transported between two conducting plates was explained. Recall that in one direction of the "E" field, energy is transferred from the "E" field to the accelerating electron. An electron traveling at high speed in the reverse direction can transfer energy from the decelerating electron to the "E" field. This latter phenomenon will be examined in more detail. It provides the basis for generating microwave power by both the klystron and the magnetron.

## 6b. The Klystron

The elementary klystron, depicted in Figure 19, is a microwave amplifier tube that makes use of two cavities of the type illustrated in Figure 18. The cross-sectional drawing shown in Figure 19 is a

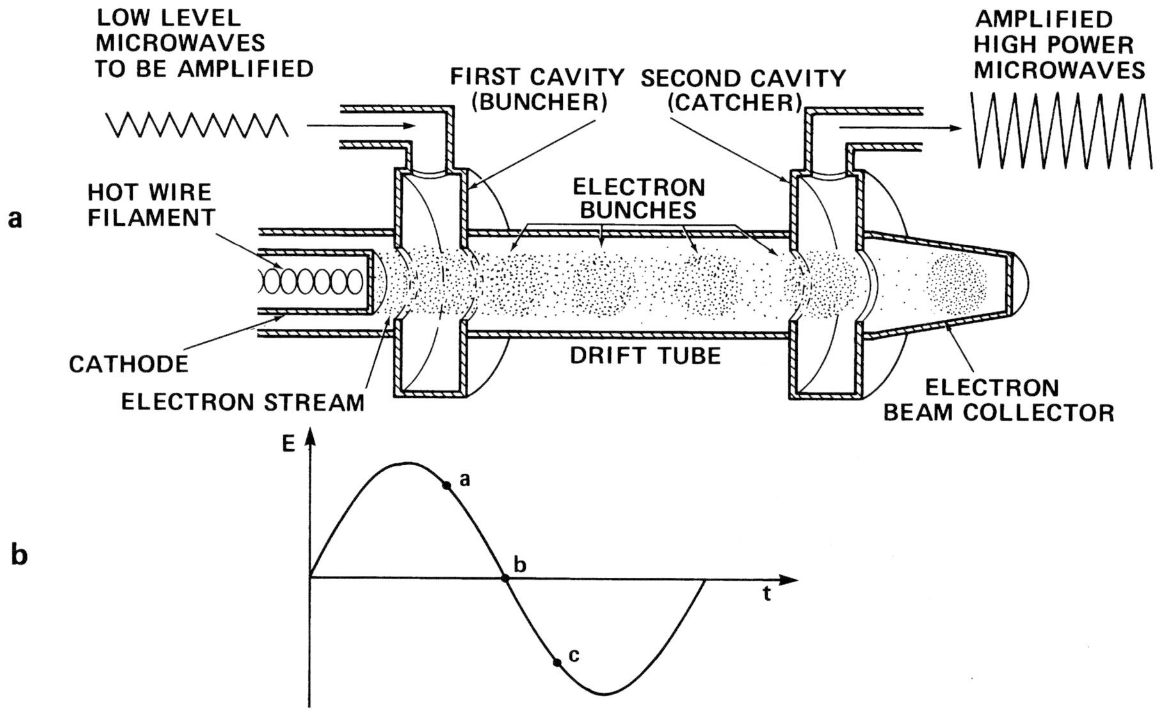


Figure 19. (a) Cross-sectional drawing of an elementary two-cavity klystron tube used as a microwave power amplifier. The two cavities are shown in cutaway sections similar to the section shown in Figure 18b. A drift tube connects the two cavities. The anode is not a single identifiable component of the klystron, but consists largely of the electron beam collector on the right end. (b) The timing diagram is for the "E" field of the first, or buncher, cavity. The "E" fields vary sinusoidally across the cavity gaps as illustrated for increasing time, t.

view that contains the cylindrical Z axis of the cavities, similar to the view shown in Figure 18b. On the left is the cathode, the source of electrons for the klystron, which is given a negative pulse of voltage. This accelerates electrons into the first, or buncher cavity, as it is called. The buncher cavity is energized by very low-power microwaves that set up alternating "E" fields across the gap between left and right cavity walls. The "E" fields vary in time, as shown in Figure 19b. Recall that it is the negative "E" field that accelerates the electrons. Those electrons that arrive early in the microwave cycle, at times between points a and b, encounter a retarding "E" field and are slowed. The velocity of those electrons arriving at time b, when the "E" field is zero, is not affected. Electrons arriving at later times, between points b and c, are speeded up by the negative "E" field. This causes the electron stream to form bunches. This process is called velocity modulation, since it alters the velocity but not the average number of electrons in the beam. They pass along the drift tube connecting the two cavities. The electrons moving with different velocities merge into discrete bunches, as shown in Figure 19.

The second, or catcher cavity, has a resonance at the arrival frequency of the bunches. As the electron bunches leave the drift tube and traverse the catcher cavity gap, they generate a retarding "E" field by inducing charges on the ends of the cavity and, thereby, initiate an energy conversion process. By this process, much of the electron's kinetic energy of motion is converted to intense "E" fields in the second cavity, creating microwave power that is used to energize the linear accelerator structure. The residual beam energy that is not converted to microwave power is dissipated as heat in the electron beam collector on the far right. The heat is removed by the water cooling system. The beam collector of high-powered klystrons is shielded with lead to attenuate hazardous x-rays created by these stopped electrons. Such klystrons have three to five cavities and are used with high-energy linacs, e.g., 18 MeV and above. The additional cavities improve high current bunching and increase microwave power amplification. They can provide a tremendous (e.g., 100,000:1) amplification of microwave power. The klystron is located in the stand, as shown in Figure 15, or in the modulator cabinet for some manufacturers.

Figure 20 illustrates a three-dimensional cutaway drawing of a high-power klystron that produces about 5 megawatts of peak power and is similar to that used for the Clinac 18. The electron gun (cathode) of the tube is at the bottom. The center section contains four amplifying cavities separated by drift tubes. The upper section consists of the water-cooled collector and output waveguide. This klystron is about 1 meter (m) in length and sits atop an oil-filled tank with its cathode-electron gun portion submerged to provide

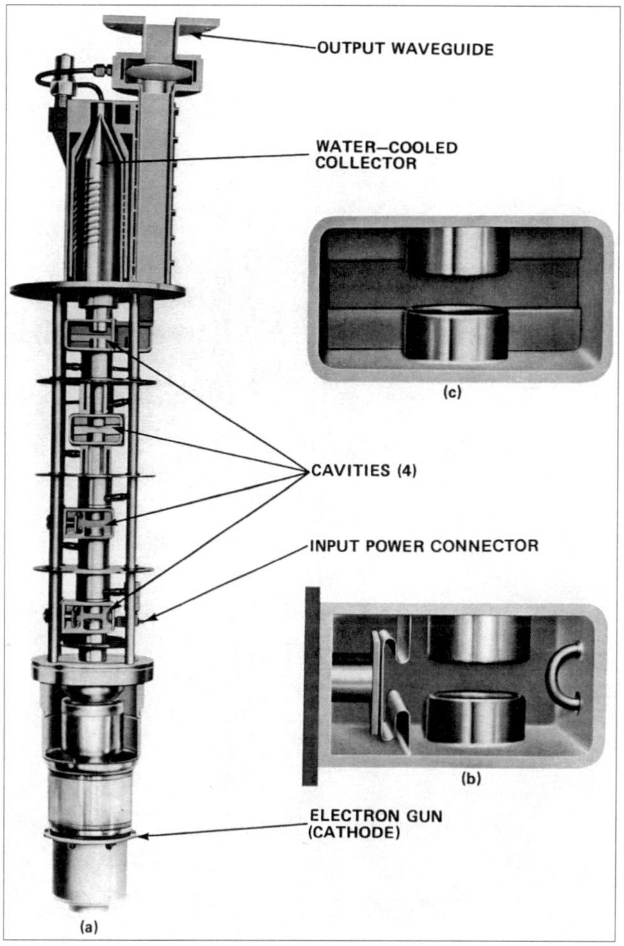


Figure 20. (a) Cutaway four-cavity klystron, similar to that employed in the Clinac 18. Views (b) and (c) are cutaway individual cavity sections. (b) Enlarged view of the bottom cavity—the input power coupling loop is on the right, and a fine-tuning device is on the left. (c) Enlarged view of cavity number three—the fine-tuning device has been cut away in this view.

the requisite electrical insulation (see Figure 15). The cathode is pulsed with a negative voltage of about 120 kV. The four cavities each have tuning adjustments (Figure 20b) that provide small changes in cavity dimensions, bringing them to the same resonant frequency of the accelerator structure operation (some klystrons are pretuned at the factory). The buncher cavity nearest the cathode (Figures 20a and b) is energized from a low-power microwave source. Cylindrical current-carrying coils, not shown here, surround the cavities and drift tubes and provide a magnetic field to confine and focus the electron beam traversing the klystron along its axis. The rectangular waveguide conducts the microwave power pulses out of the tube from the output cavity to the accelerating structure.

## 6c. The Magnetron

The magnetron is the microwave source (often called an oscillator) usually employed to power lower-energy linacs, typically 12 MeV o less, but occasionally as high as 20 MeV. Like the klystron, it is a two-element tube with a cathode and anode. Both incorporate microwave cavities. The magne-

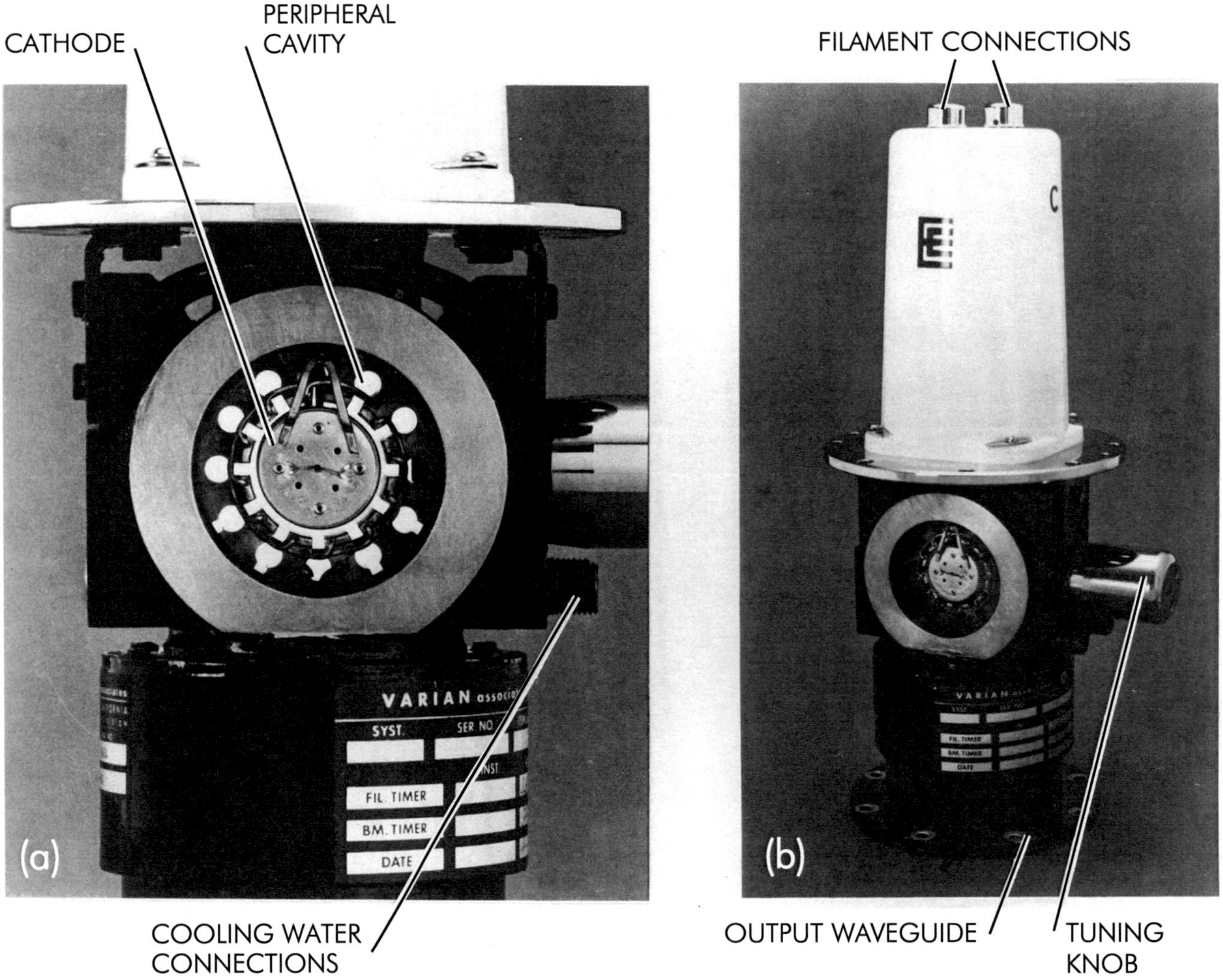


Figure 21. Cutaway magnetron of a type widely used in medical linacs. (a) The cylindrical cathode is surrounded by 12 peripheral cavities of the segmented anode. (The cavity on the top is obscured by the filament lead used for heating the cathode.) Two small coupling loops, just visible in the bottom cavity, connect the microwave power to the output waveguide just below it. The cooling water connections are on the right. (b) The two filament connections for heating the cathode are on the top; the output waveguide is on the bottom. A fine-tuning knob is on the right.

tron is usually a less costly (but a less stable) microwave power source than the klystron. The magnetron shown in Figure 21 has cylindrical geometry, shown in the circular cross section of Figure 22. The central cylindrical cathode is surrounded by the evacuated drift space and then by an outer anode having 12 cavities. The cylindrical cathode is heated by an inner filament connected to each end of the cylinder, one end of which can be seen in Figure 21a and b. Circular geometry is characteristic of the magnetron; linear geometry is characteristic of the klystron (compare Figures 19 and 20 with Figures 21 and 22).

Figure 22 is a cross section made by cutting a slice at mid-depth, parallel to the surface shown in Figure 21a. A static magnetic field, "H," is applied perpendicular to the plane of the cross section shown. In addition, a pulsed electric field, $E_p$, directed radially inward all around, is applied between the central cathode and the segmented anode that includes 12 cavities in the outer circular wall.

The electrons emitted from the cathode are accelerated by the pulsed electric field, $E_p$, toward the anode across the evacuated drift space between cathode and anode. The accelerated electrons induce an additional (+,–) charge distribution shown on the anode poles and an electric field, $E_m$, of microwave frequency between adjacent segments of the anode (see Figure 22) in a manner similar to that in the catcher cavity of the klystron. In addition, the magnetic field, "H," imparts a circular arc component to the electrons' motion. Thus, they move in complex spirals, S, under the combined influence of $E_p$; the magnetic field, "H"; and the induced microwave electric

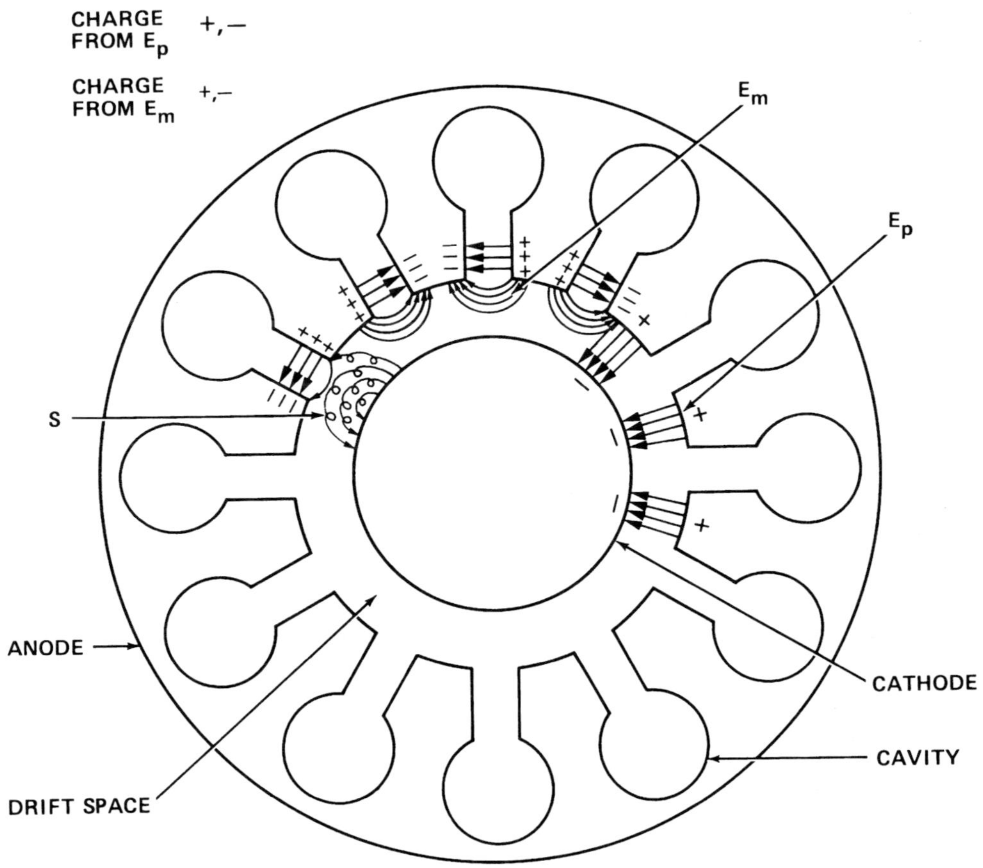


Figure 22. Cross-sectional drawing of the magnetron of Figure 21 showing representative electric fields $E_p$ (pulsed) and $E_m$ (microwave), with associated electric charge (+,–) distributions. At a particular time and place in the drift space, electrons move in typical paths, S, under the influence of the magnetic field that is perpendicular to the cross section and the sum of electric fields $E_p$ and $E_m$, which are shown separately.

field, $E_m$. In the process, approximately 60% of the kinetic energy of the electron beam is converted into microwave energy. Magnetrons almost invariably function as high-power oscillators, that is, originators of microwave power, but klystrons usually operate as amplifiers driven by a low-power oscillator. However, if one feeds back a small portion of the output of a klystron to its first cavity, it can function as an oscillator.

The power output of magnetrons and klystrons is measured in thousands of watts (kW) or millions of watts (MW). The watt is the unit of electrical power, that is, the rate at which electrical energy is expended. A household electric iron or toaster consumes about 1 kW of electrical power. Typically, magnetrons that operate at a frequency of 3000 MHz (corresponding to a 10-cm wavelength) provide 2 MW peak power output during a burst of radiation, although 4- to 5-MW versions are available at increased cost. The magnetron need only be energized for one one-thousandth of the time to provide the usual short bursts of radiation. Thus, the magnetron shown in Figure 21 operates at 2 MW peak output power and 2 kW average power output and is widely used in medical linacs. Typically, a klystron operates at 3- to 7-MW peak output power and a 3- to 7-kW average output power with an efficiency of about 50%.

# 7 THE WAVE GUIDE AND CIRCULATOR

Microwave power is conveyed from the klystron (or magnetron) to the accelerator structure by a system of hollow pipes called waveguides (Figure 15). These are either rectangular or circular in cross section, as shown in Figure 23. For example, the short, circular waveguide between the stand and gantry (Figure 15) facilitates rotation of the gantry. The short circular section is between rectangular waveguides. Waveguides replace the traditional electrical wires and cables, which are inefficient in transmitting power at microwave frequencies. Waveguides confine microwaves by reflecting them forward off the walls like a hose or pipe confines water flowing through it. They are pressurized with an insulating gas, sulfur hexa-

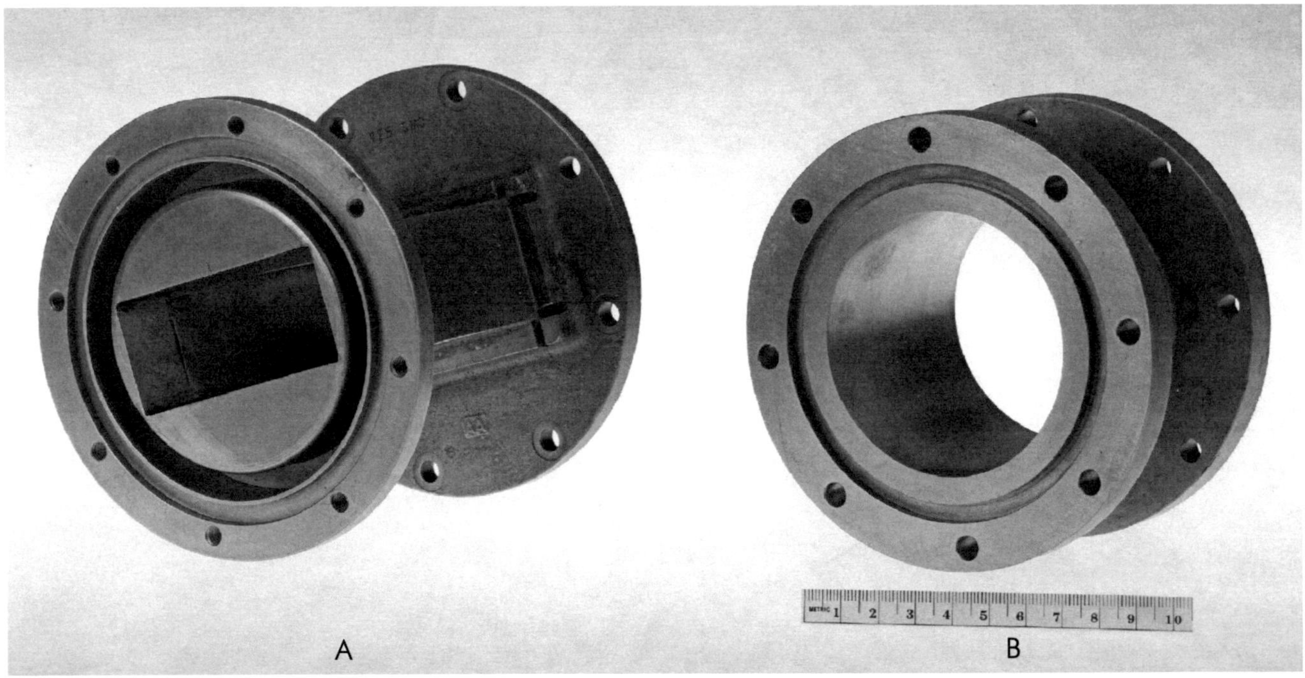


Figure 23. (A) Rectangular waveguide approximately 3.5 × 7.5 cm in cross section; (B) cylindrical waveguide approximately 7.5 cm in inside diameter. (Courtesy of Stanford Linear Accelerator Center, Stanford, CA.)

fluoride, which reduces the possibility of electrical breakdown and thus increases their power-handling capacity. Two ceramic windows separate the pressurized waveguide from the evacuated klystron at one end and from the evacuated accelerator structure at the other end. The windows are transparent to the microwaves.

The circulator (Figure 15), placed in the waveguide between the klystron and the accelerating structure, acts like a one-way street sign, permitting traffic to move forward through a street intersection, but allowing two-way traffic just beyond the intersection. Reflected microwave power is diverted aside in the circulator and absorbed, similar to the way traffic approaching the intersection from just beyond it can be diverted to a side street. Microwave power is allowed to proceed forward from the klystron and through the circulator to the accelerator, but microwave power that is reflected back from the accelerator structure is prevented from reaching the klystron (or magnetron) where it could lead to instabilities and damage.

# 8 ACCELERATOR STRUCTURES

The linac accelerator structure (sometimes called the accelerator waveguide) is a long series of adjacent, cylindrical, evacuated microwave cavities located in the gantry, as shown in Figures 15 and 42. It makes use of the cavity principles that have been discussed in Section 6a. Here, however, the objective is to transfer energy from the cavity "E" fields to an accelerating electron beam. Medical accelerator structures vary in length from 30 cm for a 4-MeV unit to more than 1 m for the high-energy units.

The first few cavities vary in size. They both accelerate and bunch the electrons in a manner like that of our klystron buncher cavity described earlier. Typically, only about one-third of the injected electrons are captured and accelerated by the microwave "E" field. As the electrons gain energy, they travel faster and faster until they reach relativistic velocity, almost the speed of light. The first few cavities are designed to propagate an "E" field with an increasing velocity in order and to further bunch and accelerate the electrons to relativistic velocity. Later cavities are uniform in size and provide a constant-velocity traveling wave at the velocity of light. Initially, electrons gain energy predominantly by increasing their velocity, later by an increase in their relativistic mass because their velocity cannot attain the speed of light. For example, a 2-MeV electron moves at 98% the speed of light. Its mass in motion is almost five times its mass at rest. Here, we are invoking Einstein's famous mass-energy equivalence concept, that is, increased energy of rapidly moving particles appears as increased mass.

Accelerator structures are of two types: traveling wave and standing wave. The "E" field patterns behave differently in these structures and are central to understanding linacs. First, we will look at traveling-wave linacs from this "E" field viewpoint.

## 8a. Traveling-wave Accelerator Structures

A hollow, cylindrical pipe—such as the waveguide used for microwave power transmission in Figure 23b—has an "E" field pattern as shown in the cross section in Figure 24a. This pattern travels one way down the pipe from the klystron (or magnetron) faster than the electrons can keep up. Hence, a cylindrical pipe is not useful for accelerating electrons.

The traveling-wave fields are slowed by "loading" the pipe with washer-like inserts called disks (Figure 24b). Now the waveguide has been transformed into a long series of resonant cavities (compare Figure 24 with Figure 18). When energized, these very high-cavity "E" fields are suitable for electron acceleration (Figure 24) along the axis.

The microwave cavities of the accelerator structures are made of copper. Copper is used because of its high heat conductivity, which improves temperature control, and because of its high electrical conductivity, which reduces power losses. The accelerator structure shown in Figure 25 consists of a series of precisely machined parts, washer-like disks sandwiched between short cylindrical sections. This sequence of disks and short cylinders is assembled on a long spindle for a particular length (energy) structure and soldered together in a furnace. The soldering material is in the form of very thin silver washers, shown at the bottom of Figure 25. These are placed between each disk and cylinder junction

Figure 24. (a) Spatial traveling-wave electric "E" field pattern at one instant of time along the axis of a smooth cylindrical waveguide. (b) Spatial traveling-wave electric "E" field pattern at one instant of time along the axis of a disk-loaded cylindrical waveguide. The direction of the electric "E" field is reversed every half wavelength, λ/2. The pattern repeats every wavelength, λ, and there are four cavities per wavelength in the disk-loaded structure. The direction of the "E" field also reverses every half cycle in time.

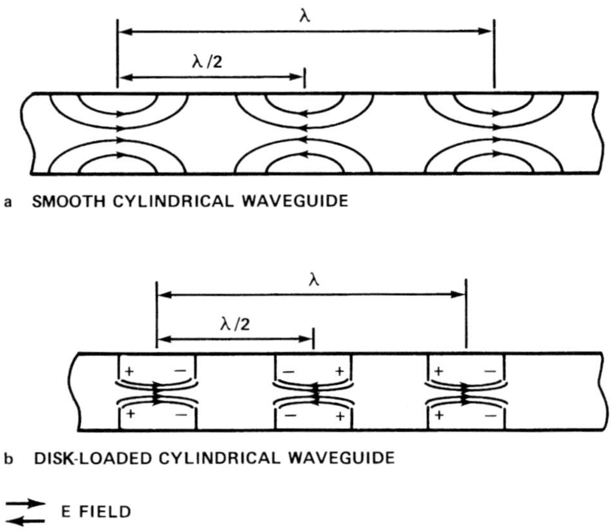

Figure 25. Components of an accelerator structure (right and bottom). Short section of an assembled accelerator structure (left). (Courtesy of Stanford Linear Accelerator Center, Stanford, CA.)

surface and, when melted, they fuse the components together. Once fused, the sections become a rigid, vacuum-tight accelerator structure. Higher energies require more cavities and longer structures.

The process is still not finished, however. The structure must be tuned to a single, precise, resonant frequency. Machining of the cavity components, prior to assembly on the spindle, is the first step in establishing the correct dimension for each cavity. This amounts to a "rough tuning" and results in a crude resonance with most cavities being "off-tune." Next, each cavity is "fine-tuned" by mechanically squeezing it to create very small dimensional changes, perhaps a few thousandths of a centimeter. Then, like a finely tuned symphony orchestra, they all play the same note. The cavities now resonate to the same frequency and provide optimal energy gain for the accelerating electrons.

As noted earlier, the electrons are captured and bunched on a moving "E" field, gaining energy by traveling in step on the advancing electric wave. At the far end the residual microwave power, not transferred to the electron beam, is absorbed by resistive material fused to the wall of the last cavity, and none is reflected. Further detail of how the wave progresses appears in Figure 26. The "E" field along the axis varies smoothly in a sine wave pattern, as shown for three sequential instants of time, and the pattern moves smoothly from left to right as time progresses. The solid arrows along the axis denote the instantaneous positions of the maximum positive (to the right) and maximum negative (to the left) values of the traveling electric wave ("E" field). Electrons are accelerated to the right on the negative portions of the "E" wave, that is, just to the right of the sine wave crest identified by arrows directed to the left. In any one cavity, the "E" field maximum reverses direction from time $t_1$ to $t_3$ (a half cycle of time) but a given wave crest (direction arrow) travels forward by one cavity from time $t_1$ to $t_2$ and again from time $t_2$ to $t_3$. From this traveling wave, an electron at a corresponding speed will gain energy in each successive cavity. Figure 27 shows an early prototype traveling-wave accelerator structure, cut in half

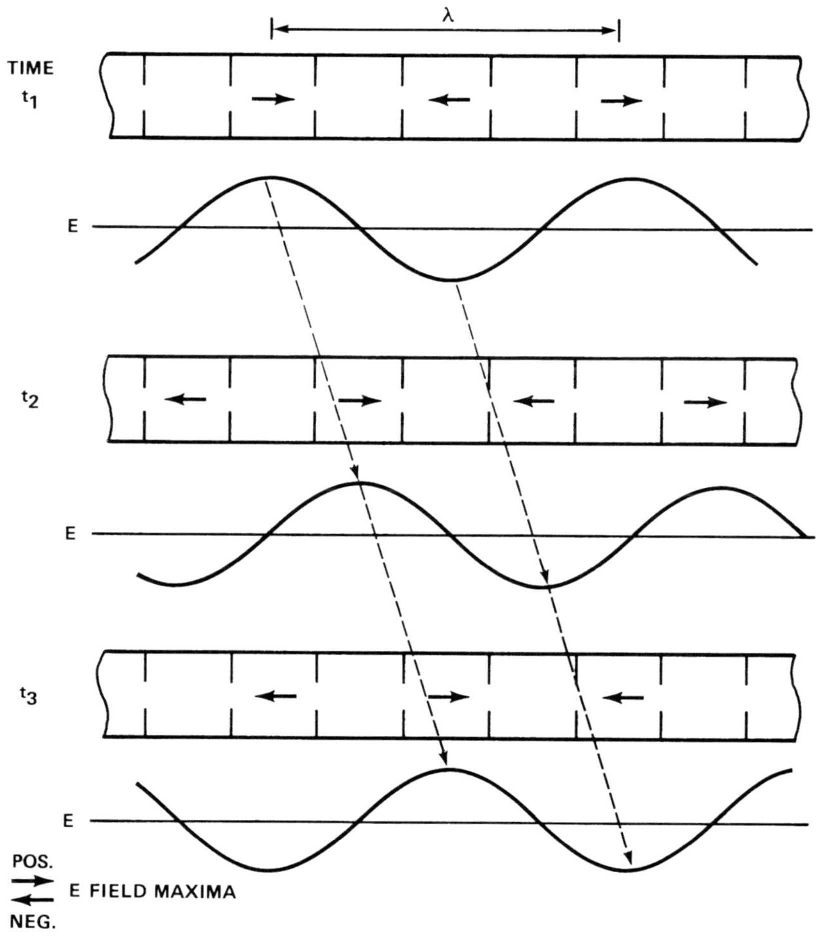


Figure 26. Illustration of a traveling-wave electric "E" field moving to the right in an accelerator structure at three sequential instants of time $t_1$, $t_2$, and $t_3$ separated by one quarter cycle in time. The dashed lines indicate how the "E" field maxima advance to the right with time. There are four cavities per wavelength, λ.

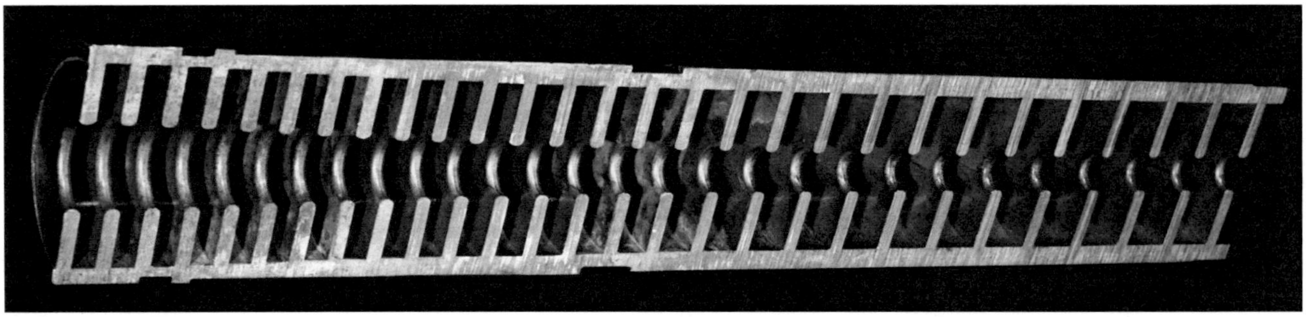

Figure 27. Cutaway traveling-wave accelerator structure; the tapered buncher section is on the left, and the uniform section is on the right.

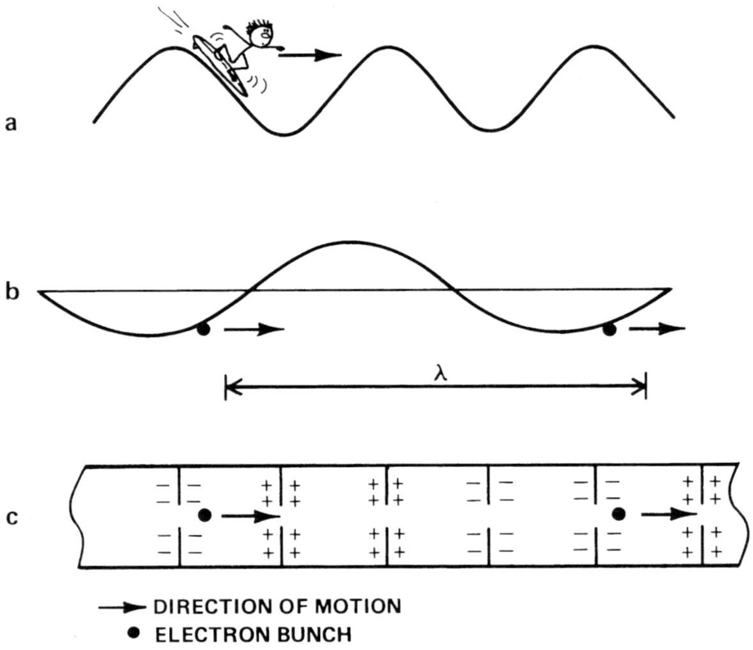


Figure 28. Traveling-wave principle: (a) for a boy surfing on a water wave advancing to the right; (b) for electrons occupying a similar position on an advancing (negative) electric "E" field; and (c) the associated charge distribution that pushes (– charge) and pulls (+ charge) the electron bunches along the cylinder. There are four cavities per wavelength, λ.

along its cylindrical axis. Note that the buncher section on the left incorporates larger and variable disk aperture sizes and more closely spaced disks than the uniform section on the right. This buncher is many cavities in length. Bunchers in contemporary linacs use fewer cavities. The input power waveguide attaches on the left (buncher) end.

A boy surfing on a water wave provides a useful traveling-wave analogy in Figure 28a. Here he is shown riding the forward edge of the crest, moving in step with the wave traveling to the right. If he slips backward over the crest of the wave, he will just bob up and down as the waves pass under him, and he will move slowly, if at all, toward the right. Similarly, electrons move forward on the front of the advancing negative "E" wave (Figures 28b and c) or are lost from it. The water particles themselves just go up and down and do not move forward, yet the top of a water wave travels forward. Similarly, the conduction electrons in the cavity walls are confined to moving back and forth between walls of a single cavity, yet the position of the "E" field maximum travels forward as a result of that movement, as shown in Figure 26.

## 8b. Standing-wave Accelerator Structures

Most present-day medical linacs are of the standing-wave type because the accelerator structure can be much shorter, and, therefore, the treatment unit is less bulky than traveling-wave linacs of comparable energy. Standing-wave linacs operate somewhat like the traveling-wave unit just described, but with one significant difference. The "E" wave varies in magnitude with time in a sinusoidal manner, but the pattern remains stationary along the axis and does not advance like the traveling "E" wave or water wave just studied. A good standing-wave analogy is the pattern of a violin string fixed at both ends and vibrating up and down to produce a musical note.

In the case of traveling-wave accelerators, microwave power is fed to the structure via the input waveguide at the proximal (electron gun) end. The residual power is absorbed at the distal (target) end of the structure. In the standing-wave accelerator, the microwave power can be fed anywhere along the length of the structure because the power proceeds in both forward and backward directions from the input waveguide and is reflected at both ends. The incident forward wave is reflected backward from the distal end, and the backward wave is reflected forward from the proximal end. There are now two waves: an advancing incident (forward) wave and a reflected (backward) wave. These two waves are reflected back and forth from one end to the other end of the accelerator structure about one hundred times during a 5-microsecond pulse. The circulator, described earlier, stops reflected power from reaching and detuning or damaging the klystron or magnetron.

Figure 29 shows the "E" field maximum values, denoted by arrows for these two waves at three sequential instants in time $t_1$, $t_2$, and $t_3$. The

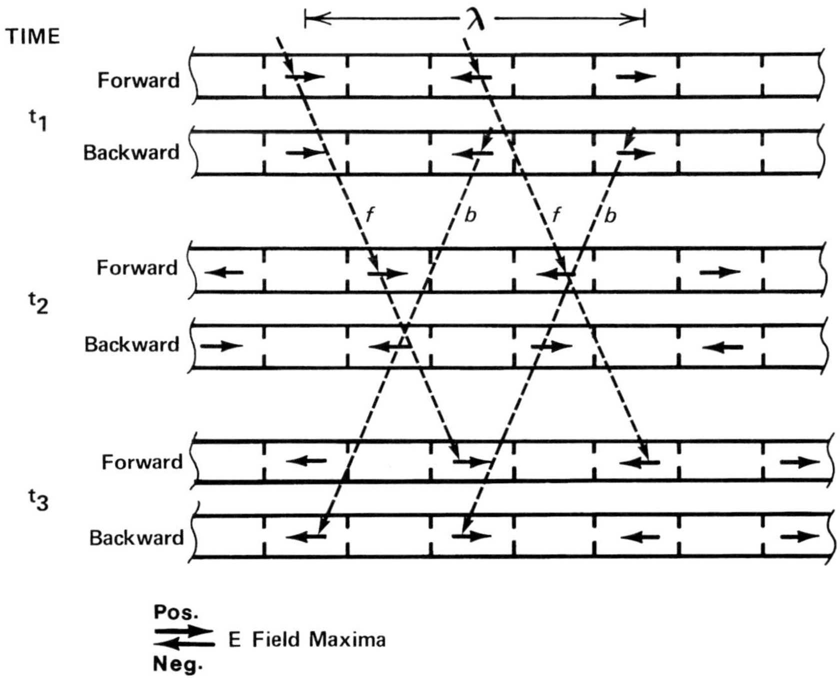


Figure 29. Standing-wave principle illustrating forward (*f* moving to the right) and backward (*b* moving to the left) "E" field maxima at three sequential instants of time. The dashed lines indicate the time sequential positions of the forward and backward moving "E" field maxima. They exist simultaneously in a standing-wave accelerator structure but are shown here in two separate structures for clarity.

forward wave crests (instantaneous positions denoted by arrows) moving to the right advance one cavity length during the time interval from $t_1$ to $t_2$, $t_2$ to $t_3$, etc. Similarly, the backward wave crests move at the same speed to the left. These sequential movements can be seen by examining each of the two patterns of arrows at the three times. Here, the sine wave "E" field patterns have been omitted and attention is confined to the wave crests denoted by arrows.

The effective "E" field, in accelerating the electron beam, is the sum of the forward and backward waves, as shown in Figure 30. Its magnitude, assuming 100% reflection and no losses, is double that of either the forward or backward wave when the fields in a single cavity are in the same direction. However, it is zero when the fields in a single cavity are in opposite directions. The effective "E" field exhibits a sinusoidal pattern with distance along the accelerator structure, as shown in Figure 30. The crests of the sine wave pattern oscillate up and down with the progression of time.

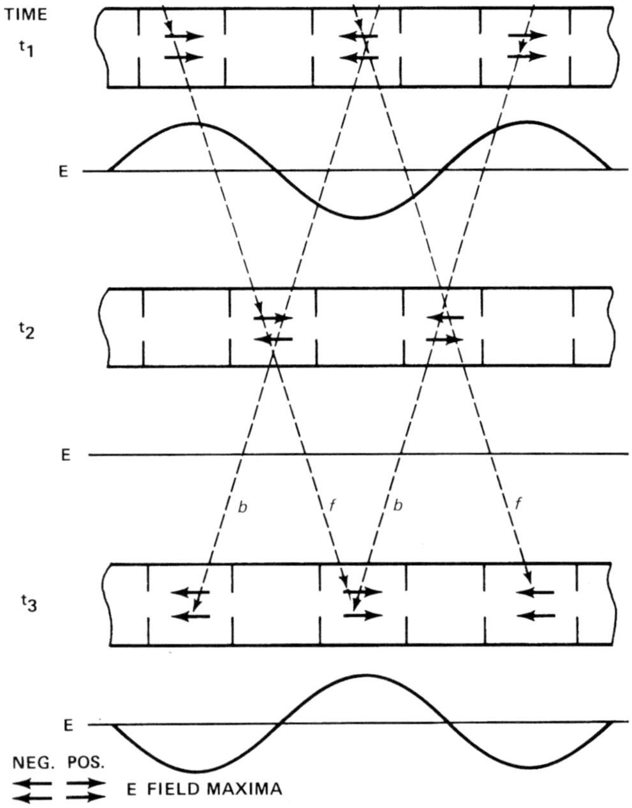


Figure 30. Standing-wave electric "E" field patterns in an accelerator structure for combined forward and backward waves at three sequential instants of time. Two traveling waves moving in opposite directions, as shown in Figure 29 (*f* forward, upper maxima; *b* backward, lower maxima), generate such a standing wave. The pattern shown at time $t_1$ will recur one half cycle later at time $t_3$ with polarity reversed.

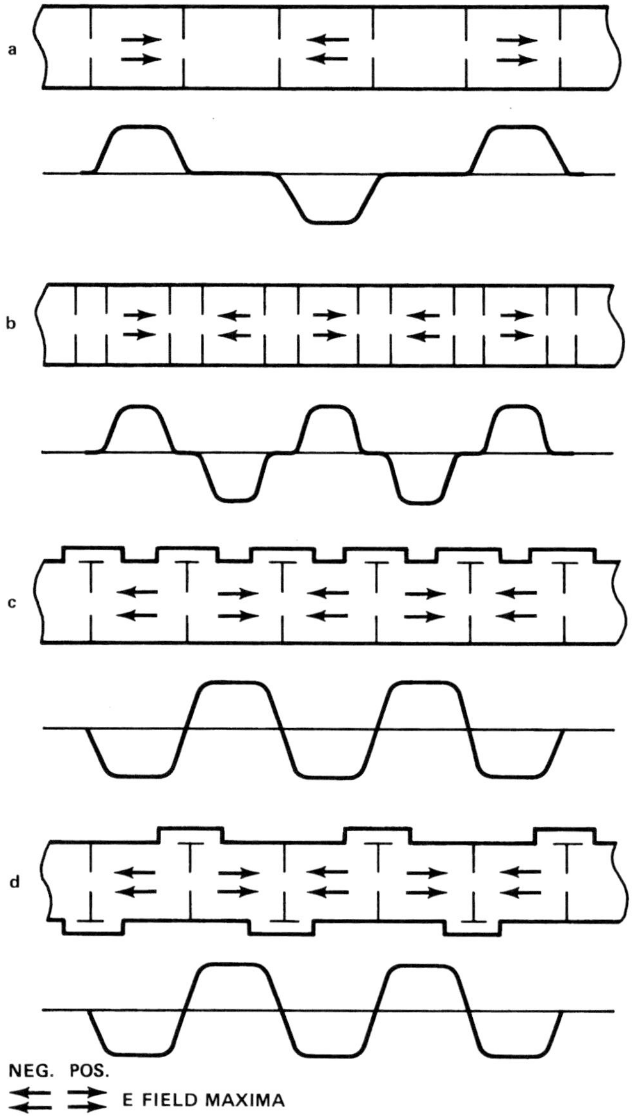


Figure 31. Evolution of side-coupled standing-wave accelerator structure from a linear standing-wave structure and related "E" wave patterns along the axis. The "E" field pattern below each structure shows the spatial field along the axis at the same time in the microwave cycle. The waves are squared due to the presence of harmonics of the fundamental frequency.

Note that every other cavity of this standing-wave structure in Figure 30 has a zero "E" field at its center at all times—at times $t_1$ and $t_3$ because both the forward and backward "E" fields are zero and at times $t_2$ because the forward and backward "E" field are equal in magnitude but opposite in direction, and cancel completely (see Figure 29). These zero "E" field cavities are essential in transporting microwave power but do not contribute to electron acceleration. Their role is to transfer or couple power between accelerating cavities. Because they play no role in acceleration, they can be moved off-axis and the length of the structure can be shortened.

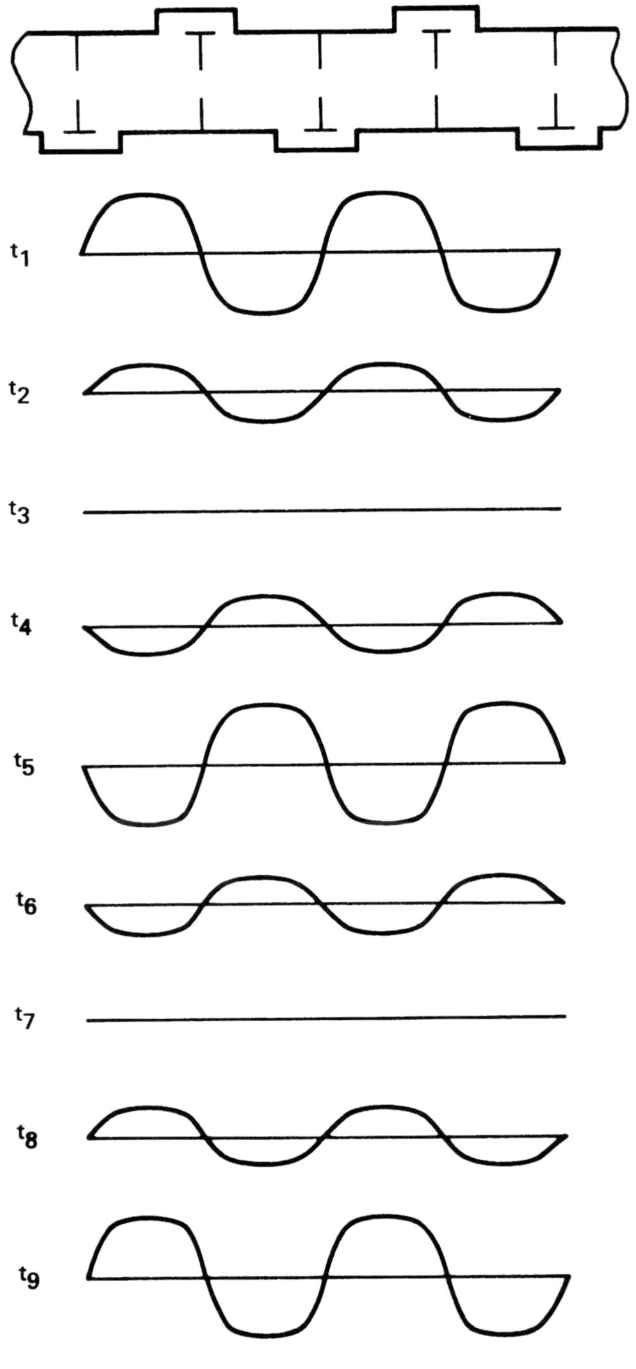


Figure 32. A sequential look at the axial standing-wave "E" field pattern for one full microwave cycle of time for the bimodal structure depicted in Figure 31d. Note that the "E" field pattern does not advance, but changes in magnitude and direction with time.

Figure 31 illustrates how the shortened, side-coupled standing-wave structure evolves from the standing-wave structure of Figure 30. First, every other cavity of Figure 31a, which couples power between accelerating cavities, is shortened in length, as in Figure 31b. The resonant frequency of a cavity depends on its diameter, not its length. Next, they are moved off-axis, as in Figure 31c, and finally, in Figure 31d, placed on alternating sides of the axial accelerating cavities. The spatial "E" field pattern shown below each sequential accelerator structure is for the same time in the microwave cycle. In Figures 26, 28b, and 28c, the "E" wave repeats every four cavities, and there are four cavities per wavelength λ. At any given instant, only one of four cavities is accelerating the electron bunch, and the other three cavities are "coasting." In Figures 31c and 31d, the "E" wave repeats every two axial cavities so that, at any instant, half of the axial cavities are accelerating the electron bunch, and the small relatively lossless off-axis coupling cavities replace half of the cavities of the traveling-wave accelerator, hence the shorter length and greater efficiency for the standing-wave design. Figure 32 illustrates in detail how the axial "E" spatial pattern changes in time over a complete microwave cycle for a standing-wave linac. Contrast the time variation of this pattern to that for a traveling-wave linac in Figure 26. Note that the "E" field pattern does not advance, but changes in magnitude and direction with time. We can now optimize the cavities along the beam axis for acceleration and the off-axis coupling cavities for microwave power transport. Figure 33 is a cutaway view of such an optimized standing-wave accelerator structure. This is called a bimodal or side-coupled accelerator structure. Two standing-wave accelerator structures constructed in this way are shown in Figures 34 and 35. They are shorter in length than a traveling-wave structure for a given energy gain and a given klystron or magnetron power.

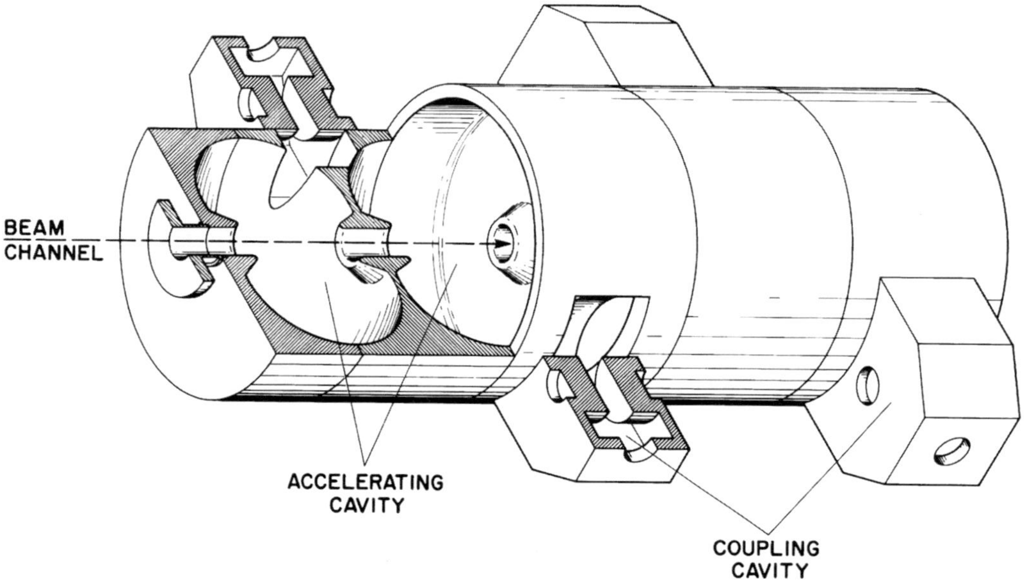


Figure 33. Cutaway view of a bimodal or side-coupled standing-wave accelerator structure. The accelerating cavities are shaped for optimum efficiency. The coupling cavities are staggered to reduce asymmetries introduced by the coupling slots. (Courtesy of Los Alamos Scientific Laboratory, Los Alamos, New Mexico.)

Electrons injected into standing-wave structures, such as those illustrated in Figures 34 and 35, are captured, bunched, and accelerated in the first few cavities, just as in the traveling-wave accelerator. They pass through the following cavities during the "E" wave's negative excursion and are accelerated (recall that electrons are accelerated in the opposite direction of "E"). During that time, the "E" wave of the next adjacent cavity is positive and electrons are not accelerated in it. However, as the electron bunch crosses the boundary between adjacent cavities, the "E" wave in the next cavity starts its negative excursion and the electron bunch is again accelerated. Each cavity accelerates electrons only when its "E" field is negative. This process continues until the electrons acquire their final energy.

# 9 MULTIPLE X-RAY ENERGY MODE ACCELERATORS

Most modern high-energy radiation therapy linacs provide multiple x-ray energy modes. For example, a machine might provide a 6-MV x-ray mode, an 18-MV x-ray mode, and an electron mode with energies of 6, 9, 12, 15, and 18 MeV. Multiple x-ray energy mode linacs are available using either a standing-wave or a traveling-wave accelerator structure. For multiple x-ray energy mode linacs, the accelerator structure length must be adequate to provide the high-energy x-ray mode at a high dose rate. This structure must also provide the low-energy x-ray mode at a high dose rate. This latter requirement necessitates a high beam current because x-ray production efficiency and, therefore, dose rate is markedly reduced at the low electron energy. All electron mode energies require much smaller beam currents, which are easily provided.

## 9a. Standing-wave Accelerators

Early linacs operated at one low x-ray energy, constrained by the use of low-power magnetrons and the need for high beam current for adequate x-ray output (e.g., 1 gray/min at 100 cm SSD) and reasonably short treatment times. Treatment beam energy is increased by higher microwave power input for a given beam current. As accelerator structures and microwave power sources became better understood, variable-energy linacs

Figure 34. Cutaway view of a standing-wave linac structure. Electrons attain an energy of 4 MeV in this 30-cm-long structure with five accelerating cavities. The input waveguide is on the bottom, the electron gun attaches on the left, and the x-ray target is permanently sealed into the structure on the right end. (Courtesy of Varian Medical Systems, Inc. All rights reserved.)

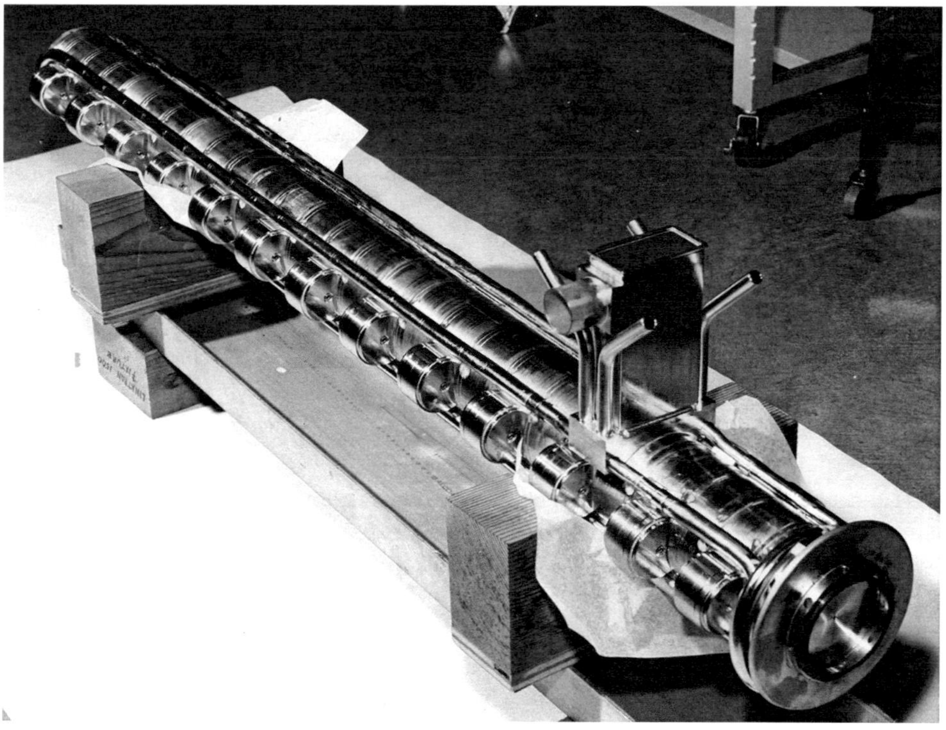

Figure 35. Linac standing-wave structure. The input waveguide is on the top, just right of center. Water-cooling tubes, which are soldered to the structure, can also be seen. (Courtesy of Varian Medical Systems, Inc. All rights reserved.)

for electron therapy and multiple-energy linacs for x-ray therapy were developed.

The standing-wave structure for a multiple-energy linac can be considered as consisting of two portions, the output end and the gun end. If, in order to reduce x-ray beam energy (e.g., from 18 to 6 MV), the magnitude of the accelerating "E" field is reduced in the second portion (output end) of an unmodified standing-wave accelerator structure, the magnitude of the "E" field drops correspondingly in the first portion (gun end). Similarly, if the phase of the sine wave "E" field is shifted in the second portion, it shifts equally in the first portion. Because the RF power is reflected back and forth in a standing-wave structure, the first portion senses and adjusts to the field in the second portion, and vice versa.

If the magnitude of the "E" field is correct for 18 MV operation—that is, optimal capture, bunching, and positioning of electrons injected from the electron gun on the "E" wave crest and for acceleration through the remainder of the structure—then the magnitude of the "E" field must be reduced for 6-MeV operation. A large beam current must be attained so as to provide a high 6-MV dose rate. In present multiple-energy linacs there are two fundamentally different ways to modify standing-wave accelerator structures to eliminate or reduce this problem of inadequate electron capture, bunching, and positioning in the low x-ray energy mode.

1. *Change the ratio of microwave power fed to the first and second portions of the standing-wave accelerator structure.* This can be done by using a compact energy switch in a side cavity located between the first and second portions, as shown in Figure 36. In one position of the energy switch, the side cavity provides high coupling between the first and second portions, creating a single high-magnitude "E" field in both portions of the accelerator structure for the high x-ray energy mode. In the second position of the energy switch, the side cavity provides low coupling, creating a low-magnitude "E" field in the second portion of the accelerator structure for the low x-ray energy mode. In both positions of the energy switch, the required "E" field is maintained in the first portion of the accelerator structure in order to maintain optimal capture and bunching of the injected electrons from the gun and positioning the resulting bunch near the crest of the "E" wave. This results in an electron beam of stable energy with a narrow energy spread and, hence, high transmission through the bending magnet and minimal leakage radiation due to electrons lost from the beam before reaching the x-ray target. This provides high dose rate and a stable x-ray beam in both low- and high-energy modes. The energy switch employs one moving part, a plunger. Alternately, instead of an energy switch, a high-power microwave circuit can be used, employing a power divider and a phase shifter. Such systems, which necessitate modification of the design and operation of the structures, are bulky and employ many moving parts.
2. *Use a broad band buncher in a standing-wave accelerator structure.* In the first portion of such an accelerator structure (the buncher), the cavities are made very short. Also, the coupling slots to the side cavities may be small to reduce the magnitude of the "E" field in this first portion. There is no energy switch, so the magnitude of the "E" field is one value for high-energy x-ray mode and a much lower value for low-energy x-ray mode, throughout the accelerator structure. This is achieved by reducing the output of the klystron or magnetron for the lower energy mode.

Because the initial cavities are so fore-shortened, the electrons injected from the gun are captured and bunched around a position very far forward of the crest of the accelerating "E" field sine wave. Following this bunching section there is one especially long cavity. In passing through this cavity, the electron bunch slips backward relative to the "E" field sine wave to near (but not on) its crest, for acceleration through the rest of the accelerator structure; ahead of the crest in a high "E" field for high x-ray energy mode; and behind the crest in a low "E" field for low x-ray energy mode.

This technique avoids the use of a mechanically moving part, namely the plunger in the energy switch. However, it is wasteful of RF power, requiring a higher-power klystron or magnetron, and the off-crest acceleration produces an output beam with larger energy spread and greater energy instabilities. This makes it more difficult to obtain high transmission of the electron beam through the bending magnet to ensure a high dose rate flattened fully to the corners of

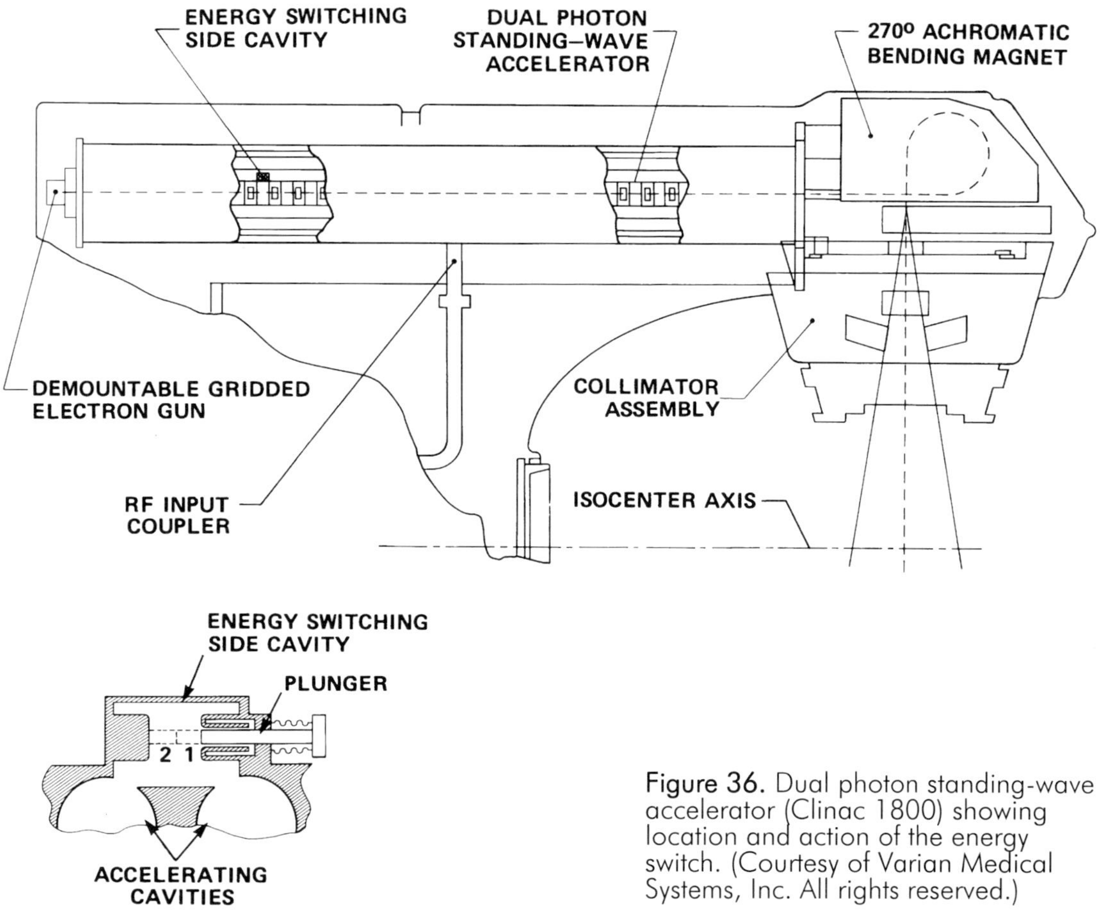


Figure 36. Dual photon standing-wave accelerator (Clinac 1800) showing location and action of the energy switch. (Courtesy of Varian Medical Systems, Inc. All rights reserved.)

large fields in low-energy x-ray mode, with stable dose distribution over all gantry angles.

## 9b. Traveling-wave Accelerators

In traveling-wave accelerator structures without RF feedback through an external circuit, the first portion does not sense the field in the second portion because the wave travels only forward. The amplitude of the accelerating "E" field can be changed in the second portion of the accelerator structure without a significant effect on the capture and bunching properties of the first portion. One way of producing a downward taper of this "E" field from first to second portions is by beam loading, simply increasing injected beam current from the gun and keeping the klystron or magnetron power constant. Because the RF power is being transferred to the high-current electron beam, a progressively decreasing fraction of the RF power flowing through the accelerator structure is left to produce the "E" field in the cavities of the second portion. Also, the phase of the "E" field sine wave can be tapered from the first portion to the second portion, simply by changing the frequency of the klystron or magnetron; the electron bunch then slips in phase over the "E" field sine wave, receiving less than maximal acceleration.

As pointed out in Section 8b, traveling-wave structures are much longer than standing-wave structures for the same input RF power, beam energy, and beam current. Such long accelerator structures for multiple x-ray mode accelerators can be accommodated more easily in a drum-type gantry. Here the accelerator structure can project through the barrel-like drum, which rotates on bearings mounted on the floor.

External RF feedback from the second portion output to the first portion input is used for some traveling-wave structures in order to improve frequency stability (so that a magnetron can be used in a high-energy accelerator), and in this respect

the first portion senses the field in the second portion. Variable coupling is used in such feedback circuits to maintain the same accelerating "E" field in the first portion, both in high x-ray energy mode with light beam loading and in low x-ray energy mode with heavy beam loading.

The frequency and beam stability of traveling-wave structures and magnetrons are inherently less stable than standing-wave structures and klystrons. Linac designs incorporating them rely more heavily on electronic feedback, such as using computer look-up tables, to maintain treatment beam stability.

# 10 BENDING MAGNET

The electron beam leaving the accelerator structure continues through an evacuated bending magnet system. Here it is deflected magnetically so as to either strike a target for x-ray therapy or to exit through the treatment head, via a thin metallic window, for electron therapy. The exit axes coincide for x-ray therapy and for electron therapy. Note the location of the bending magnet in Figure 15.

The bending magnet deflects the beam in a loop of approximately 270° (Figures 37 and 38). This magnet design provides focusing for the spread of energies in the beam. Whereas a simple 90° deflection magnet will defocus and spread the

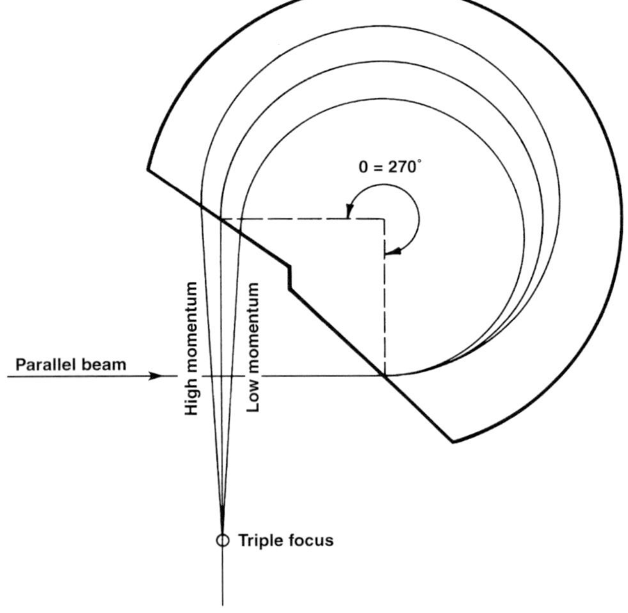


Figure 37. A simplified 270° beam-bending magnet (singly achromatic) with focusing properties as shown. The magnetic field, H, is perpendicular to the plane of the electron orbits. A full doubly achromatic magnet includes additional angular and spatial focusing properties not shown here but described in the text. For example, it also provides transverse focus, that is, focusing in a plane at right angles to that shown. (Courtesy of *Physics and Medicine in Biology*, Vol. 18, pp. 321–54, 1973 and C. J. Karzmark, Ph.D.)

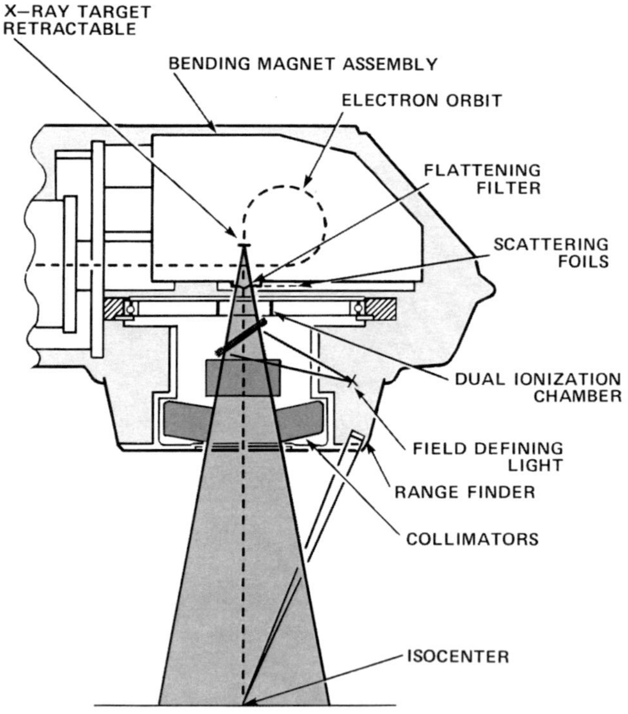


Figure 38. Treatment head of a Clinac 18. (Courtesy of Varian Medical Systems, Inc. All rights reserved.)

beam, a small x-ray focal spot helps ensure that the x-ray treatment fields have sharply defined edges (i.e., a small penumbra), a feature that is of assistance in treatment. This feature improves uniformity of radiation of the tumor and spares nearby critical organs. Medium- and high-energy accelerators employ bending magnets. However, many low-energy units have a straight-through beam without a bending magnet. This is because these accelerator structures are short enough to be vertically mounted and still allow isocentric rotation up to 6 MV.

As shown in Figure 37, the lower-energy electrons are deflected through a loop of smaller radius, and the higher-energy electrons are deflected through a loop of larger radius. The important property of this 270° achromatic magnet is that these components of energy are brought back together to the same position, angle, and beam cross section at the target, as they were when they left the accelerator structure. Thus, a 3-mm diameter beam out of the accelerator is reproduced as a 3-mm diameter beam at the target. If a simple 90° deflection magnet were to be used, electrons with different energies would exit with slightly different trajectories, analogous to the separation of different wavelengths of white light by a triangular prism.

However, in the singly achromatic magnet of Figure 37, a variation in mean beam energy will result in a change in the angle of the beam at the target, producing angular x-ray field asymmetry, even though the focal point position stays fixed. To correct for this angular dependence on energy, modern magnets are doubly achromatic. The mean energy of the beam can vary without changing the mean position or angle of the beam at the x-ray target, hence, maintaining symmetry of the treatment field.

# 11 TREATMENT HEAD

The **treatment head** (Figure 38) contains a number of beam-shaping, localizing, and monitoring devices. The high-energy x-rays emerging from the target are forward-peaked in intensity, being of higher intensity along the beam central axis and of progressively less intensity away from it (see Figure 39). If desired, the forward-peaked x-ray lobe is flattened to provide uniform treatment fields. This is accomplished by the **flattening filter**, a conical metal absorber, placed on the axis as shown. It absorbs more photons from the intense central axis and fewer from the periphery of the beam. Many treatment machines also have a flattening-filter free (FFF) mode, in which the flattening filter is not used in order to increase the achieved dose rate, especially near the central axis of the beam. In FFF mode, a relatively thin beam-hardening filter is rotated in the beam path to remove low-energy photons and electrons from the beam.

The **dual ionization chamber** system samples the radiation beam (x-rays or electrons) passing through the treatment head and produces electrical signals that terminate the treatment when the prescribed dose is given. Two independent ion-

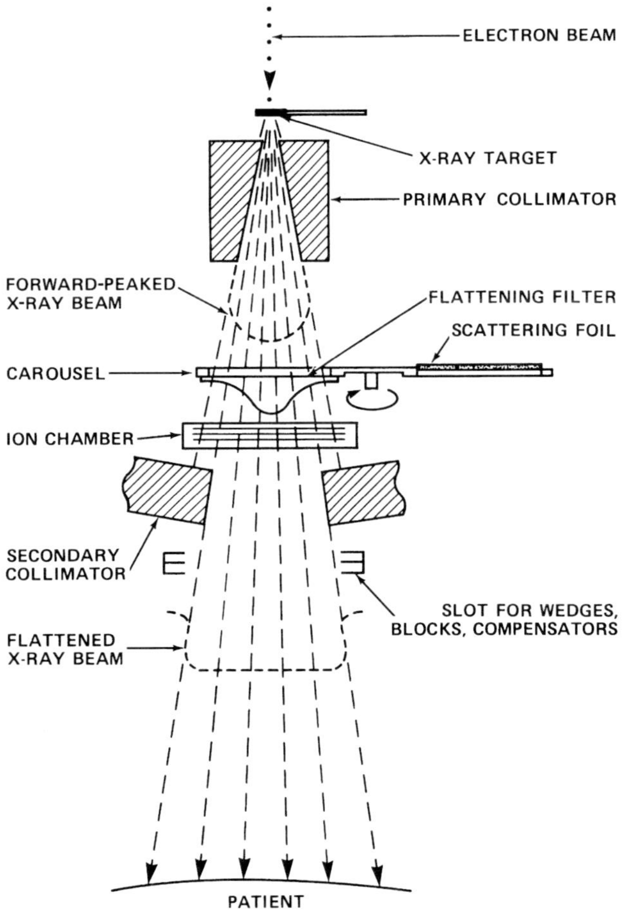


Figure 39. Beam subsystem for x-ray beam therapy.

ization chamber channels ensure that the prescribed dose is delivered accurately and safely, with one serving as a check on the other. The ionization chambers may be segmented such that asymmetries in the beam can be detected and used to interlock beam generation.

The **field-defining light** simulates the x-ray field by illuminating the area to be irradiated on the patient's skin surface and facilitates positioning the patient for x-ray treatment. It provides an intense light field, duplicating in size and shape the x-ray field incident on the patient as defined by the collimators or other beam-limiting devices such as beam blocks or apertures.

A **range finder** (optical distance indicator) light projects a numerical scale on the patient's skin to indicate the source-skin distance (SSD) from 80 to 130 cm on the central axis.

The **x-ray target** and **field flattener** or beam-hardening filter are retractable and are moved off-axis for electron therapy.

Additional details of the treatment head subsystem for **x-ray therapy** are shown in Figure 39. A **primary collimator** limits the maximum field size for x-ray therapy. The effect of the flattening filter on beam uniformity is shown. The forward-peaked x-ray beam has been flattened. Treatment field size is defined by the **secondary collimator** consisting of four thick metal blocks, called jaws, often made of tungsten. To help provide sharp edges for treatment fields, the jaws are focused to the beam source: the movement of the blocks is confined to arcs so that the block faces present a flat edge parallel to the beam diverging from the target. They are adjustable in pairs (or independently in some units to provide asymmetric

Figure 40. Multi-leaf collimator (MLC). Note the irregularly shaped field defined by the MLC leaves. (Courtesy of Varian Medical Systems, Inc. All rights reserved.)

fields) and, in some linacs, provide rectangular treatment fields as large as 40 × 40 cm at a distance of 1 m from the target. The secondary collimator rotates about the beam axis, allowing angulation of fields.

Individual patient tumors vary in size and shape. **Multi-leaf collimators (MLCs)** (Figure 40) allow radiation beams to conform to the profile of the tumor from the beam's eye view. MLCs consist of a pair of jaws divided into thin leaves, each typically 0.25–1.0 cm wide projected to isocenter. The position of each leaf is computer controlled. MLCs may replace the upper jaw, the lower jaw, both jaws, or be located as a tertiary collimator below the jaws. Leaf edges in the direction perpendicular to leaf motion are focused to the beam source. In a "double-focused" MLC, the leaf ends in the direction of motion are also focused to the beam source, requiring the leaves to move in arcs. Single-focused MLCs have linear leaf motion and use rounded leaf edges to achieve approximately constant transmission penumbra across the entire range of leaf motion. MLCs are typically constructed of tungsten alloy because of its high density. The MLC design shown in Figure 40 has 52 motorized leaves, each 1 cm wide at isocenter, with 16 cm travel across the beam centerline. MLCs greatly facilitate modern therapy techniques, such as **3D conformal therapy** and **intensity-modulated radiotherapy** (IMRT).

Accessories to modify the emergent x-ray field externally—such as wedges, tissue compensators, individually shaped apertures, and shadow blocks—may be mounted on trays that slide into slots of an accessory mount attached to the treatment head. In many systems, the accessory tray "reads" the code of the accessory to assure that the planned accessory is in place when required for a treatment field.

Additional details of the treatment head subsystem for **electron beam therapy** are shown in Figure 41. The x-ray target is moved out of the beam and a thin **scattering foil** replaces the flattening filter. A rotating carousel or sliding tray(s) facilitates this exchange. The scattering foil spreads out the small, pencil-like beam of electrons and provides a flat, uniform electron treatment field. For electron therapy, a **detachable electron applicator** is attached to the **accessory mount** of the treatment head. Final field definition is provided by a removable aperture located at the end of the applicator close to or in contact with the patient's skin. In addition, the secondary x-ray collimator is set to a field size somewhat larger than that defined by the applicator.

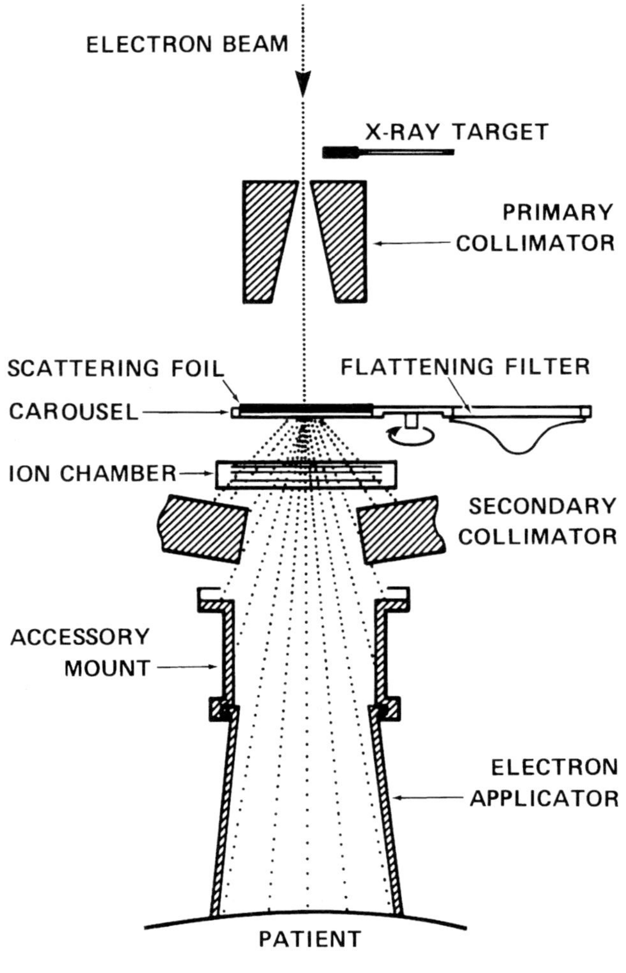


Figure 41. Boom subsystem for electron beam therapy.

# 12 IMAGE-GUIDED RADIOTHERAPY

In many modern radiotherapy techniques, treatment fields are designed to tightly conform to the lesion, minimizing the amount of healthy tissue irradiated. In such cases, it is of utmost importance that the patient is in the planned position relative to the treatment beams. In image-guided radiotherapy (IGRT), patient positioning is verified and refined using imaging performed inside the treatment vault with the patient in treatment position. Commonly, x-ray imaging is performed using either the treatment beam or a dedicated kilovoltage (kV) x-ray generator similar to those employed in diagnostic x-ray imaging. Typically, kV images are formed using flat-panel detectors. An **electronic portal imaging device** (EPID) is located in line with the treatment beam, downstream of the patient, and forms images using the treatment beam.

In a common configuration, the kV x-ray source and flat-panel detector are mounted at 90° rotation relative to the orientation of the treatment beam (as in Figure 1). In addition to forming planar radiographs, both the kV and MV systems can generate volumetric images called **cone-beam computed tomography** (CBCT) scans, by full or partial rotation about the patient.

In some systems, a stereoscopic pair of x-ray generators are mounted in the floor adjacent to the treatment machine, with corresponding ceiling-mounted flat-panel detectors. This configuration facilitates image acquisition during noncoplanar beam delivery, but this also limits the imaging field of view due to much greater source-to-imager distance.

# 13 PHYSIOLOGIC BEAM GATING

In modern, highly conformal radiotherapy, the presence of internal lesion motion due to respiration or digestive processes increases the required **treatment margin** (region in space about the tumor which is targeted with radiation) which, in turn, increases the amount of normal tissue irradiated. One approach to reducing or eliminating internal motion margins is **gating**, a technique in which the treatment beam is held when the tumor moves outside a predetermined location. Essential elements of gating include a) a mechanism to measure lesion motion, either directly or via an external surrogate, and b) a mechanism to "hold" beam delivery when the target is outside the treatment field. If an external surrogate is used, a method of correlating the surrogate with internal target motion is required. A critical metric of gating performance is the latency, or time between a when the target moves into or out of the treatment window and the time when the beam is turned on or held. Gating latency includes contributions from acquisition of the target location or surrogate signal, processing of the signal to determine the gating decision, signal travel times between subsystems, and time needed to enable or hold the beam after the linear accelerator receives the gating decision. Predictive models may be used to mitigate these latencies. Net latencies in the range of 5 to 500 milliseconds are typical of a variety of clinical gating systems.

A similar overall work flow, as follows, is common to several gating systems. Thoracic surface position is used as a respiratory surrogate: for example, a block with infrared reflective markers placed on the patient's chest or abdomen and monitored with a dedicated tracking camera. Alternatively, direct measurement of the patient surface can be performed using 3D optical tracking. The correlation between internal tumor position and abdominal surface position is established using a respiratory-correlated computed tomography (4D-CT) scan acquired with the same type of surface tracking system as will be used for treatment. The correlation is verified at treatment time using fluoroscopy performed with the onboard kV imager or respiratory-gated kV planar x-ray images. In a Varian linear accelerator, beam hold is achieved by applying appropriate bias to the grid element of the triode electron gun. A predictive filter is used, which halts the beam in case of deviations between predicted and actual marker positions, as for example would occur during coughs or other drastic changes in breathing patterns.

# 14 FUNCTIONAL BLOCK DIAGRAM AND AUXILIARY SYSTEMS

A number of auxiliary systems are essential for operation, control, and monitoring of the linac treatment unit. These systems, together with the major components described earlier, are shown in Figure 42, a functional block diagram. The modulator cabinet (which may include the klystron) and control console, shown on the left, are located outside the treatment room; the stand, gantry, and treatment couch, shown on the right, are inside. The modulator cabinet may be placed inside the treatment room in some installations.

The modulator cabinet contains a pulsed power supply, as shown in Figure 42, which energizes the klystron and the electron gun when triggered by a timing pulse (Figure 43a) from the console. The pulsed power supply provides a a 120-kV pulse of approximately 5 microseconds duration to the klystron, which generates the

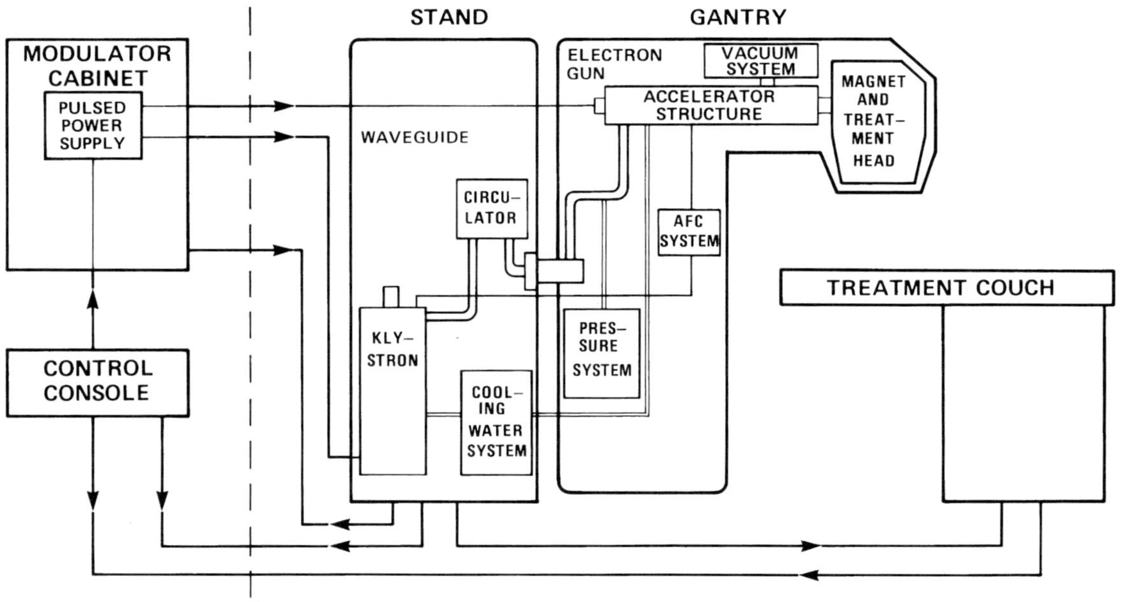


Figure 42. Block diagram of a high-energy, bent-beam medical linac (Clinac 18). Major components, auxiliary systems, and interconnections are identified. (Courtesy of Varian Medical Systems, Inc. All rights reserved.)

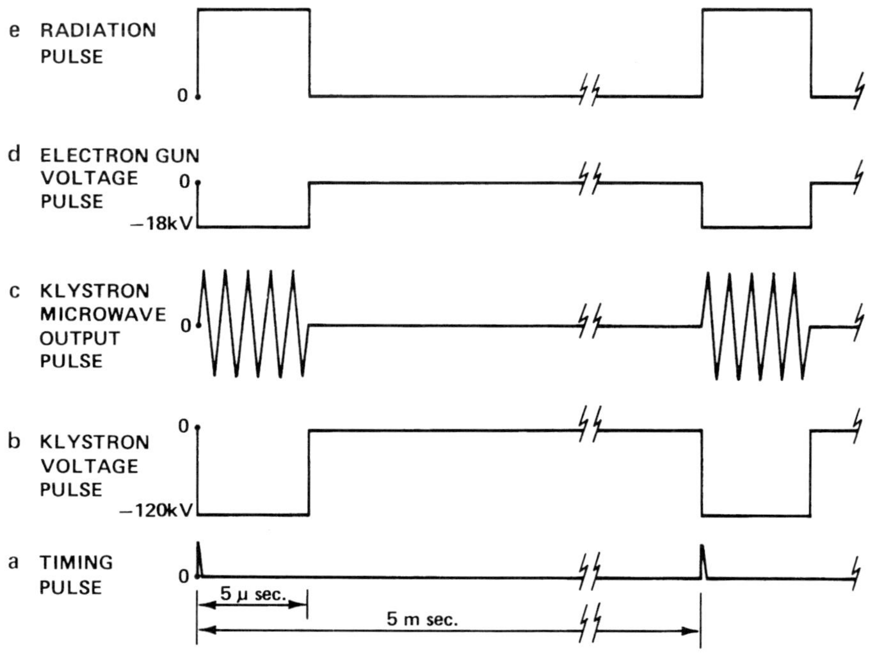


Figure 43. Timing diagram showing major time/event relationships for two sequential bursts of radiation. Note the two different abscissa time scales, one in microseconds (μ sec) and the other in milliseconds (msec).

microwave power, and a similar 18-kV pulse, which speeds electrons from the electron gun into the accelerator structure (Figures 42b, c, and d). The timing pulse rate provides a convenient method of varying the linac output dose rate.

Electrons are injected into the structure on axis from the electron gun, as shown in the upper left of the gantry in Figures 15 and 42. The gun is pulsed with a negative 18-kV pulse. As a result, electrons enter the cavities with about 18 keV of energy and a velocity approaching one-fourth the speed of light.

The **vacuum system** provides the extremely low pressures needed for operation of the electron gun, accelerator structure, and bending magnet system. Without a vacuum, the electron gun would rapidly "burn out" like a light bulb filament exposed to air. In addition, the accelerated electrons would collide with air molecules, deflecting them and reducing their energy, and the small, pencil-like beam of electrons would be broken up and diffused. The vacuum is maintained by an **electronic ion pump**. It was this latter development, more than any other, that transformed the linac from a laboratory instrument into a practical clinical tool. Earlier accelerator vacuum systems involved oil-based rotary and diffusion pumps that required significant maintenance.

The **pressure system** pressurizes the waveguide with sulfur hexafluoride, an insulating gas. This is needed to prevent electrical breakdown from the high-power microwave electric "E" fields.

A **cooling system** providing temperature-controlled water establishes the operating temperature of sensitive components and operates primarily to remove residual heat dissipated in other components.

**Temperature control** is particularly critical for the accelerator structure itself. Without it, the series of cavities composing the accelerator structure can change dimensions slightly. The effect of this is to "detune" them in the same way a musical instrument changes its pitch; they are then "off-frequency," and their acceleration capability is seriously impaired.

An **automatic frequency control (AFC)** system continuously senses the optimum operating frequency of the accelerator structure to maximize radiation output. It uses this information to "tune" the klystron or magnetron to this microwave frequency.

An elaborate **monitor and control system** maintains control of linac operation and patient treatment. It monitors operation to ensure proper linac performance and to ensure that the prescribed treatment is faithfully delivered in a safe manner. Deviations, depending on their nature and magnitude, will give rise to fault warning signals or termination of the treatment, when appropriate. The center of this monitor and control function is at the **control console** with connections to all other units. The control console provides status information on treatment modality, accessories in use, set dose, dose delivered, interlock status, emergency off, as well as other data pertinent to linac operation and patient treatment. Frequently, the monitor function is directly linked to the control function, and current status information is used in a feedback manner to maintain optimal performance.

A multitude of quantitative and procedural checks are incorporated in the console to ensure correct, safe operation. The digital logic circuits used in modern computers are the basis for these checking procedures. They can be carried out in a few seconds and are assessed automatically, prior to each treatment.

A counting system, tied to the dose monitor, terminates the treatment when the set dose **monitor units (MU)** are delivered. A backup timer is set to terminate treatment in the event of dose monitor system failure. The radiation therapist observes the treatment on a monitor (see Figure 4), and a two-way audio system permits communication with the patient.

# 15 OPERATIONAL REVIEW

An extensive treatment plan is prepared prior to treatment for each individual patient, usually consisting of a sequence of treatments carried out over a period of one or more weeks. The radiation oncologist specifies the tumor location and volume, treatment fields, and the prescribed dose for the patient's tumor. The number of monitor units to deliver one unit of absorbed dose in gray (Gy) varies for field size, modality, energy, and treatment distance. It is customary to define a dose monitor unit as the ionization produced under defined conditions. Generally, the linac is adjusted so that 1 (1.00) MU equals 1 (1.00) centigray at the depth of dose maximum ($d_{max}$) on the central axis in water for a 10 cm × 10 cm field at an SSD of 100 cm for a given energy and modality. For example, 150 MU would deliver 1.5 Gy for these conditions. The dose elsewhere is specified by isodose curves (actually surfaces), that is, curves of constant dose that are arbitrarily set to 100% where passing through $d_{max}$. Isodose curves are generated on patient computed tomography (CT) scans during treatment planning using model-based dose calculation methods or Monte Carlo calculations.

At the beginning of a treatment session, the radiation therapist loads the treatment plan on the control console. The treatment plan contains gantry, collimator and table angles, multi-leaf collimator parameters (if required), monitor units, field sizes, and beam modalities and energies for each treatment beam. The patient is positioned on the treatment couch by the radiation therapist. Accessory beam-modifying devices, such as blocks or wedges, are attached and positioned. The radiation therapist then proceeds to the control console. Image guidance in the form of on-board x-ray, electronic portal imaging devices, or other techniques may be used to verify and refine the patient's positioning. Images may be viewed in real-time on dedicated consoles (see Figure 4). The therapist then sets the controls to deliver the daily dose. In typical medical linacs, electrons used directly for treatment have several energies from about 3 to 25 MeV. However, a particular medical linac may have as few as only one or two x-ray energies, or as many as up to five x-ray energies, anywhere from 4 to 25 MeV.

Before the treatment begins, an internal check system is automatically activated that sequentially verifies linac operating parameters for correct values. In most units, a method for testing the dosimetry system is used to ensure that the prescribed dose will be delivered. The treatment system may also include a computer-based **record and verify system**, which compares the treatment that has been set up with a record of the intended treatment (see Figure 4). This treatment assessment may include field size and collimator angulation, gantry angle, couch position, the daily dose for each field, and the preset monitor readings to provide that dose. Such record and verify systems identify setup errors prior to treatment so that they can be corrected when they exceed a preset magnitude, for example, more than one degree of arc. Such programs verify and record each treated field on a continuing, daily basis throughout the course of treatment. Medical accelerator safety and quality control considerations are reviewed by the reports of Task Groups 35, 40, 45, and 142 of the American Association of Physicists in Medicine (AAPM). The report of AAPM Task Group 100 provides a methodology for systematic and comprehensive quality management in radiation therapy.

Typically, the linac is pulsed several hundred times per second, with the exposure for each treatment field lasting from a few seconds to a few minutes. When the "beam on" button is pushed, an elaborate sequence is initiated, in part described by the timing diagram of Figure 43. First, the modulator accumulates energy for the first pulse of radiation. It sends out two high-voltage pulses in unison: one to give the electrons leaving the electron gun their first boost of energy as they enter the accelerating structure (Figure 43d), the other to energize the klystron (Figure 43b). The klystron then delivers the microwave power to the accelerating structure (Figure 43c) and, in turn, to the electron beam emerging from the electron gun. Here, the intense "E" fields come into play, bunching the electrons and accelerating them to their final energy.

The electron beam next traverses the bending magnet and is directed on the x-ray target, or scattering foil, in the case of electron therapy. In the case of x-ray beams, a flattening filter may be used to make the x-ray intensity uniform over the entire field area. The emerging cone of radiation traverses the two monitor ionization chambers and is further shaped by the collimator and other beam-shaping devices.

Figure 43 summarizes pertinent time relationships for two sequential bursts of radiation. In this diagram of idealized timing, the linac is pulsed every 5 milliseconds, that is, 200 times per second. The timing pulse that initiates each sequence is very short, and all other pulses are of about 5 microseconds duration. During this 5-microsecond interval, 15,000 complete micro-wave cycles occur (3,000 / μsec × 5). This microstructure is also present in the radiation burst (Figure 43e), but timing details have been omitted for simplicity.

X-ray treatment beams can be verified by comparing exit fluence, measured using the electronic portal imaging device, with estimated fluences based on the treatment plan, a process referred to as portal dosimetry.

The authors hope that this primer leads to a better understanding of medical linear accelerators, promotes clear communication between all personnel involved, and contributes to the safe and effective treatment of patients.

# BIBLIOGRAPHY

Karzmark, C. J. (1984). "Advances in Linear Accelerator Design for Radiotherapy." *Med. Phys.* 11(2):105–28.

Karzmark, C. J., C. S. Nunan, and E. Tanabe. *Medical Electron Accelerators.* New York: McGraw-Hill, Inc., 1993.

Karzmark, C. J. and N. C. Pering. (1973). "Electron Linear Accelerators for Radiation Therapy: History, Principles and Contemporary Developments." *Phys. Med. Biol.* 18:321–54.

Kramer, S., N. Suntharalingam, and G. F. Zinninger, Eds. *High Energy Photons and Electrons: Proceedings of an International Symposium on the Clinical Usefulness of High-Energy Photons and Electrons (6-45 MeV) in Cancer Management.* Thomas Jefferson University, Philadelphia, PA, May 22–24, 1976. New York: John Wiley and Sons, 1976.

Purdy, J. A. et al. (1993). Medical Accelerator Safety Considerations. Report of AAPM Radiation Therapy Committee, Task Group #35. *Med. Phys.* 20(4):1261–75.

Tapley, Norah duV, Ed. *Clinical Applications of the Electron Beam.* New York: John Wiley and Sons, 1976.